ULTRAVIOLET DISINFECTION FOR WASTEWATER— LOW-DOSE APPLICATION GUIDANCE FOR SECONDARY AND TERTIARY DISCHARGES

2015

Water Environment Federation
601 Wythe Street
Alexandria, VA 22314–1994 USA
http://www.wef.org

International Ultraviolet Association (IUVA)
1718 M Street, N.W. #276
Washington, DC 20036
http://www.iuva.org

Ultraviolet Disinfection for Wastewater—Low-Dose Application
Guidance for Secondary and Tertiary Discharges

Copyright © 2015 by the Water Environment Federation and International Ultraviolet Association. All Rights Reserved. Permission to copy must be obtained from WEF and IUVA.

Water Environment Research, WEF, WEFTEC, and Water's Worth It are registered trademarks of the Water Environment Federation.

ISBN 978-1-57278-312-6

IMPORTANT NOTICE

The material presented in this publication has been prepared in accordance with generally recognized engineering principles and practices and is for general information only. This information should not be used without first securing competent advice with respect to its suitability for any general or specific application.

The contents of this publication are not intended to be a standard of the Water Environment Federation® (WEF) or the International Ultraviolet Association (IUVA) and are not intended for use as a reference in purchase specifications, contracts, regulations, statutes, or any other legal document.

No reference made in this publication to any specific method, product, process, or service constitutes or implies an endorsement, recommendation, or warranty thereof by WEF or IUVA.

WEF and IUVA make no representation or warranty of any kind, whether expressed or implied, concerning the accuracy, product, or process discussed in this publication and assume no liability.

Anyone using this information assumes all liability arising from such use, including but not limited to infringement of any patent or patents.

About WEF

Founded in 1928, the Water Environment Federation (WEF) is a not-for-profit technical and educational organization of 36,000 individual members and 75 affiliated Member Associations representing water quality professionals around the world. WEF members, Member Associations, and staff proudly work to achieve our mission to provide bold leadership, champion innovation, connect water professionals, and leverage knowledge to support clean and safe water worldwide. To learn more, visit www.wef.org.

About IUVA

The International Ultraviolet Association (IUVA) is a non-profit organization with over 600 members representing UV professionals around the world. The vision of the organization is to make the use of ultraviolet technology a leading technology for public health and environmental applications, and to position IUVA as the leading authority on the use of ultraviolet technology through advocacy to the education, industry, research and public policy sectors worldwide. To learn more about the IUVA, visit www.iuva.org.

Prepared by the **Ultraviolet Disinfection for Wastewater** Task Force of the **Water Environment Federation**

Katherine (Kati) Y. Bell, Ph.D., P.E., BCEE, *Chair*

Archis Ambulkar, M.S.
Robert Bastian
Jeffrey L. Bertacchi, EE, M.P.A.
Ernest (Chip) R. Blatchley III, Ph.D., P.E., BCEE, F. ASCE
Christian Bokermann, Dr.-Ing.
Keith Bourgeous Ph.D., P.E.
Cynthia L. Bratz, P.E., LEED AP
Joseph Cantwell, P.E.
S. Rao Chitikela, Ph,D., P.E., P.Eng., BCEE
James Clifton
Ivan A. Cooper, P.E., BCEE
Matt Crow, P.E.
Bradford Derrick, P.E.
Gil Dichter, BA, MBA
Bertrand W. Dussert, Ph.D.
Sidney Ellner
Ufuk G. Erdal, Ph.D., P.E.
Hans X. Figueroa, P.E.
Nicola Fontaine
Val Frenkel
John M. Friel, P.E.
Ronald Gehr, Ph.D., P.Eng.
Joshua E. Goldman, Ph.D.
Lola Guerra, Chemist MSc
Clara Haenchen
Mike Harmer, P.E., C.P.E.
Rhonda E. Harris. P.E.
Gary L. Hunter, P.E., BCEE
David L. Jaeger
Samuel S. Jeyanayagam, Ph.D., P.E., BCEE
Stéphane Jousset, P.E.
Anjana Kadava, P.E.

Justyna Kempa-Teper, Ph.D., P.Eng.
Ken Kershner
Dale E. Kocarek, P.E., BCEE
Jeremy Kraemer, Ph.D., P.Eng.
Oliver Lawal
Dennis Lindeke
Karl G. Linden, Ph.D.
Jorj Long
Hadas Mamane, Ph.D.
Melanie A. Mann, P.E.
Mike McGehee, P.E.
Rick McIntyre
Sharon A. McMillin, Ph.D.
Michael Minkebige
Devon A. Morgan
Jerry A. Morgan
Michael Newbigging, P.Eng., M. Eng.
Chad Newton, P.E.
James J. Newton, P.E., BCEE, ENV SP
Morayo Noibi
Sanath Palipana
Malar Perinpanayagam, P.E., EnvSP
Peter Petersen
Andrew Salveson, P.E.
Karl Scheible
Vamsi Seeta, P.E., BCEE, LEED AP
Youngwoo (Young) Seo, Ph.D.
Michael Sevener, P.E., BCEE
Chengyue Shen, Ph.D., P.E.
Bill Sotirakos
Robert B. Stallings, P.E.
Jay Swift, P.E.
Daniel L. Theobald
Bree Trembly

Ron Trygar, CET
Joseph Viciere, D.A.S., P.E., BCEE
Michel Wanna, PMP
G. Elliott Whitby, Ph.D.

David Winters, P.E.
Wayne Wong, M.A.Sc., P.Eng., PMP
Paul Wood, P.E.
Yu Yang, Ph.D, EIT

Under the Direction of the **Disinfection Subcommittee** of the **Technical Practice Committee**

2015

Water Environment Federation
601 Wythe Street
Alexandria, VA 22314-1994 USA
http://www.wef.org

International Ultraviolet Association (IUVA)
1718 M Street, N.W. #276
Washington, DC 20036
http://www.iuva.org

Special Publications of the Water Environment Federation

The WEF Technical Practice Committee (formerly the Committee on Sewage and Industrial Wastes Practice of the Federation of Sewage and Industrial Wastes Associations) was created by the Federation Board of Control on October 11, 1941. The primary function of the Committee is to originate and produce, through appropriate subcommittees, special publications dealing with technical aspects of the broad interests of the Federation. These publications are intended to provide background information through a review of technical practices and detailed procedures that research and experience have shown to be functional and practical.

Water Environment Federation Technical Practice Committee Control Group

Eric Rothstein, C.P.A., *Chair*
D. Medina, *Vice-Chair*
Jeanette Brown, P.E., BCEE, D. WRE, F.WEF, *Past Chair*

V. D'Amato
M. Beezhold
Katherine (Kati) Y. Bell, Ph.D., P.E., BCEE
J. Davis
C. DeBarbadillo
S. Fitzgerald
S. Gluck
Betty J. Green
M. Hines
B. Jones
R. Lagrange

J. Loudon
S. Metzler
J. Miller
James J. Newton, P.E., BCEE, ENV SP
C. Peot
R. Pope
R. Porter
L. Pugh
J. Reeves
A. Salveson
S. Schwartz
Andrew R. Shaw, P.E.
R. Tsuchihashi
J.E. Welp
N. Wheatley

Contents

List of Figures	xvii
List of Tables	xix
Preface	xxi

Chapter 1 Introduction   1
Robert Bastian and
Katherine (Kati) Y. Bell, Ph.D., P.E., BCEE

1.0	PURPOSE		1
	1.1	Perspective	2
	1.2	History of Ultraviolet Disinfection	3
	1.3	Disinfection Criteria	4
		1.3.1 Indicator Bacteria	5
		1.3.2 Enumeration Methods for Indicator Bacteria	8
2.0	REGULATORY CONSIDERATIONS		9
	2.1	Regulatory Frameworks in the United States and Canada	9
	2.2	Regulatory Drivers for Ultraviolet Disinfection	10
3.0	OTHER RELATED GUIDANCE		11
4.0	ORGANIZATION OF THE PUBLICATION		13
5.0	REFERENCES		13

Chapter 2 Ultraviolet Disinfection Process Concepts and Equipment Systems   17
Karl G. Linden, Ph.D., and Hadas Mamane, Ph.D.

1.0	PRINCIPLES OF ULTRAVIOLET DISINFECTION		18
	1.1	Electromagnetic Spectrum	18
	1.2	Properties of Ultraviolet Light	19
		1.2.1 Properties of a Photon	19
		1.2.2 Ultraviolet Absorbance and Transmittance	20
		1.2.3 Laws of Photochemistry	21
		1.2.4 Ultraviolet Scattering	22

	1.3		Germicidal Action of Ultraviolet Light	22
		1.3.1	Nucleic Acid Damage	23
		1.3.2	Deoxyribonucleic Acid and Protein Absorbance	24
		1.3.3	Ultraviolet Action Spectrum	25
			1.3.3.1 Means to Obtain an Action Spectrum	25
			1.3.3.2 Action Spectrum by Organism	26
			1.3.3.3 Can the Deoxyribonucleic Acid Absorbance Spectrum Represent the Action Spectrum?	26
	1.4		Microbial Repair and Regrowth	27
		1.4.1	Photo-Reactivation	28
		1.4.2	Dark Repair	29
		1.4.3	Regrowth	29
2.0	ULTRAVIOLET DOSE RESPONSES OF PATHOGENS AND SURROGATES			30
	2.1		Collimated Beam Testing	30
		2.1.1	Types of Collimated Beams	30
		2.1.2	Ultraviolet Dose Measurements and Calculation	30
		2.1.3	Factors Affecting Ultraviolet Dose Calculation	32
	2.2		Data on Microbe Dose Response	33
		2.2.1	Bacteria	34
		2.2.2	Viruses	34
		2.2.3	Protozoa	35
3.0	ULTRAVIOLET LAMP TECHNOLOGIES			35
	3.1		Low-Pressure Ultraviolet Lamps	36
	3.2		Low-Pressure High-Output Ultraviolet Lamps	37
	3.3		Medium-Pressure Ultraviolet Lamps	37
	3.4		Alternative Lamp Technologies	37
		3.4.1	Ultraviolet Light-Emitting Diodes	38
		3.4.2	Microwave Ultraviolet Radiation	38
		3.4.3	Pulsed Ultraviolet Lamps	39
		3.4.4	Excimer Lamps	40
4.0	REFERENCES			40

Chapter 3		Bioassay Methods to Determine the Ultraviolet Dose (Fluence) Delivery of an Ultraviolet System *G. Elliott Whitby, Ph.D., and Bill Sotirakos*	51
1.0	INTRODUCTION		52
2.0	BIOASSAY PROTOCOL FOR WASTEWATER		53
	2.1	Planning and Preparation	54
		2.1.1 Test Ultraviolet System Characteristics	54
		2.1.2 Challenge Microorganisms Used in Validations	55
		2.1.3 Water Source Key Characteristics	57
		2.1.4 Absorbing Chemical	57
		2.1.5 Mixing and Sampling	58
		2.1.6 Lamp Variability and Ultraviolet Sensor Port Window Testing	58
		2.1.7 Measurement Equipment	58
	2.2	Inlet/Outlet Structures	58
	2.3	Test Lamps	59
	2.4	Test Conditions and Quality Assurance/Quality Control Samples	59
	2.5	Third-Party Oversight	60
3.0	MICROBIOLOGICAL TESTING		61
	3.1	Preparing the Challenge Microorganism	61
	3.2	Verifying Ultraviolet Reactor Properties and Ultraviolet-Intensity Sensor Performance	61
	3.3	Measuring Ultraviolet Dose (Fluence) Delivery	61
	3.4	Collimated Beam Testing	62
	3.5	Validation and Data Analysis	63
4.0	EXISTING DATA		63
5.0	REPORTING		64
6.0	BIOASSAY VALIDATION EXEMPLAR		65
	6.1	Validation Study	65
	6.2	Hydraulic Characterization	66
	6.3	System Parameters	67
		6.3.1 Power Measurements	67
		6.3.2 Ultraviolet Sensor Readings	67
		6.3.3 Headloss and Water Level	69

	6.4	Bioassay Testing	70
		6.4.1 Collimated Beam Analysis	71
		6.4.2 Bioassay Test Procedure	72
		6.4.3 Log Inactivation (Log I) Equation	73
		6.4.4 Bioassay Results and Data Analysis	74
6.5	SUMMARY		76
7.0	REFERENCES		76

Chapter 4 — Innovations and Advances in Ultraviolet Reactor Analysis and Validation — 79
Ernest (Chip) R. Blatchley III, Ph.D., P.E., BCEE, F. ASCE; Karl Scheible; and Chengyue Shen, Ph.D., P.E.

1.0	INTRODUCTION	79
2.0	FACTORS AFFECTING ULTRAVIOLET DISINFECTION REACTOR PERFORMANCE	81
3.0	CONTEMPORARY METHODS FOR ULTRAVIOLET REACTOR VALIDATION	83
4.0	EMERGING STRATEGY FOR REACTOR VALIDATION: STOCHASTIC APPROACH	91
5.0	REFERENCES	92

Chapter 5 — Process Design and System Sizing — 95
Andrew Salveson, P.E.; Keith Bourgeous, Ph.D., P.E.; Nicola Fontaine; Norayo Noibi; and Bill Sotirakos

1.0	DISINFECTION PERMIT REQUIREMENTS			96
2.0	WASTEWATER QUALITY EFFECTS ON ULTRAVIOLET DISINFECTION			97
	2.1	Effluent Water Quality		97
		2.1.1 Dissolved Constituents and Their Effect on Ultraviolet Absorbance and Transmittance		98
			2.1.1.1 Dissolved Organic Matter	98
			2.1.1.2 Inorganic Compounds	99
		2.1.2 Particles		100
	2.2	Water Quality Characterization Tools		101
		2.2.1 Turbidity		103

		2.2.2	Total Suspended Solids	103
		2.2.3	Ultraviolet Transmittance	106
			2.2.3.1	Seasonal and Diurnal Ultraviolet Transmittance Variability	106
			2.2.3.2	Secondary Process Effects on Ultraviolet Transmittance	106
			2.2.3.3	Chemical Effects on Ultraviolet Transmittance	107
			2.2.3.4	Industrial Effects on Ultraviolet Transmittance	108
			2.2.3.5	Sidestream Flow Effects on Ultraviolet Transmittance	109
	2.3	Upstream Processes to Improve Water Quality		110
	2.4	Fouling of Lamp Sleeves, Lamp Racks, and Channels		113
3.0	REACTOR SELECTION CRITERIA			114
	3.1	System Configurations		114
		3.1.1	Closed Vessel	114
		3.1.2	Open Channel	115
		3.1.3	Lamp Orientation	115
		3.1.4	Non-Submerged Ultraviolet Lamp Systems	115
		3.1.5	Lamp Spacing	115
	3.2	Establishing Design Criteria		116
		3.2.1	Flow	116
		3.2.2	Headloss and Water Level	117
		3.2.3	Influent and Effluent Water Quality	118
	3.3	Design Dose and Dose Control Strategies		118
		3.3.1	System Monitoring	119
		3.3.2	Dose Delivery Strategies	119
			3.3.2.1	Dose Pacing Based on Flow, Ultraviolet Transmittance, and Power Setting	120
			3.3.2.2	Dose as a Function of Flow, Ultraviolet Transmittance, and Ultraviolet Intensity	120
4.0	ULTRAVIOLET SYSTEM SIZING EXAMPLE			120
5.0	REFERENCES			124

Chapter 6		Equipment Selection, Facility Design, and Project Delivery *Katherine (Kati) Y. Bell, Ph.D., P.E., BCEE, and Joshua E. Goldman, Ph.D.*	129
1.0	INTRODUCTION		130
2.0	LIFE CYCLE COST ANALYSIS		130
	2.1	Project Capital Costs	131
		2.1.1 Ultraviolet Equipment Capital	131
		2.1.2 Construction Costs	132
	2.2	Operation and Maintenance Costs	133
		2.2.1 Power Consumption and System Efficiency	133
		2.2.2 Replacement Parts	134
		2.2.2.1 Lamps and Sleeves	134
		2.2.2.2 Ballasts and Drivers	135
		2.2.3 Cleaning Components	136
		2.2.4 Intensity Sensors and Ultraviolet Transmittance Analyzers	136
		2.2.5 Operations and Maintenance Labor	136
	2.3	Calculating Life Cycle Costs	137
	2.4	Non-Cost Considerations	140
		2.4.1 Headloss Effects	140
		2.4.2 Constructability and Maintenance of Facility Operations during Construction	141
		2.4.3 Operation and Maintenance Considerations	142
		2.4.4 Warranties, Service, and Manufacturer Reliability	142
		2.4.5 Reference Installations and Other Considerations	143
3.0	FACILITY DESIGN CONSIDERATIONS		143
	3.1	Site and System Hydraulics	143
		3.1.1 Available Head	144
		3.1.2 Open-Channel Versus Closed-Channel Systems	144
		3.1.3 Flow Splitting, Flow Distribution, and Flow Control	145
		3.1.4 Flow Measurement	145
		3.1.5 Level Control	146

	3.2	Power Requirements and Power Redundancy	146
	3.3	Process Redundancy	147
	3.4	Ultraviolet System Layout	148
		3.4.1 Lifting Devices	149
		3.4.2 Ultraviolet System Ballast Cabinets, Control Panels, and System Instrumentation	150
		3.4.3 Sleeve Cleaning Methods and Ancillary Facilities	151
4.0	REFERENCES		152

Chapter 7 Ultraviolet Project Delivery, Startup, and Commissioning — 153
Gary L. Hunter, P.E., BCEE

1.0	INTRODUCTION		154
2.0	PROJECT DELIVERY METHODS		154
	2.1	Design/Bid/Build	155
	2.2	Construction Management at Risk	155
	2.3	Design/Build	156
	2.4	Equipment Procurement Methods	156
3.0	KEY CONSIDERATIONS DURING CONSTRUCTION		158
4.0	KEY ACTIVITIES DURING STARTUP		161
	4.1	Commissioning, Startup, and Testing Plan	161
	4.2	Startup Checks	162
	4.3	Functional Acceptance Testing	162
5.0	KEY ACTIVITIES DURING PERFORMANCE TESTING		163
	5.1	Operational Acceptance Testing	163
	5.2	Performance Testing	163
		5.2.1 Testing Protocols	163
		5.2.2 Testing Duration	164
		5.2.3 Flow	164
		5.2.4 Water Quality	164
		5.2.5 Analysis of Data	165
	5.3	Alternative Performance Testing Methods	165
		5.3.1 Stress Testing	165
		5.3.2 Velocity Profile Testing	165

6.0	PERFORMANCE STANDARDS AND SYSTEM ACCEPTANCE		166
	6.1	Power Consumption	166
	6.2	Headloss	166
	6.3	Enforcing Manufacturer Guarantees	166
7.0	REFERENCE		166

Chapter 8 Operational Considerations 167
Jay Swift, P.E., and Chad Newton

1.0	ROUTINE OPERATION		168
2.0	SAFETY		168
	2.1	Electrical Hazards	169
	2.2	Ultraviolet Radiation Hazards	169
	2.3	Lifting Hazards	170
	2.4	Chemical Hazards	170
	2.5	Other Hazards	171
3.0	PROCESS MONITORING		171
	3.1	Ultraviolet-Intensity Sensors	171
	3.2	Ultraviolet Transmittance Meters	173
	3.3	Flow Monitoring	173
	3.4	Level Monitoring	174
	3.5	Temperature Monitoring	175
	3.6	Warnings and Alarms	175
4.0	OPERATIONAL STRATEGIES		175
	4.1	Energy Conservation	175
	4.2	System Redundancy (Multiple Banks and Channels)	177
5.0	ELECTRICAL AND CONTROL SYSTEMS		177
	5.1	Power Requirements and Power Redundancy	177
	5.3	Control Cabinets/Programmable Logic Controllers	178
	5.4	Supervisory Control and Data Acquisition System Integration	178
6.0	MAINTENANCE CONSIDERATIONS		180
	6.1	Lamp Replacement	180
	6.2	Sleeve Replacement	182

	6.3	Ballast Replacement	182
	6.4	Sleeve Cleaning	182
	6.5	Channel Cleaning	183
7.0	ANCILLARY SYSTEMS		184
	7.1	Chemical Cleaning Tanks/Stations	184
	7.2	Air Compressors/Blowers	185
	7.3	Bank/Module Lifting	185
8.0	TROUBLESHOOTING		185
	8.1	Electrical Issues	185
	8.2	Low Ultraviolet Transmittance	185
	8.3	Hydraulics	187
	8.4	Upstream Process Effects	187
9.0	REFERENCES		188

Chapter 9 Case Studies 191
Oliver Lawal and Bree Trembly

1.0	OVERVIEW	191
2.0	OPEN CHANNEL, HORIZONTAL ORIENTATION	191
3.0	OPEN CHANNEL, NONHORIZONTAL ORIENTATION	197
4.0	CLOSED VESSEL	201
5.0	REFERENCE	204

Index 205

List of Figures

2.1	Ultraviolet light in the electromagnetic spectrum	18
2.2	The DNA absorbance of *P. aeruginosa* and *E. coli*.	24
2.3	The UV action spectra for viral plaque-forming ability for various viruses relative to 254 nm.	27
2.4	Types of collimated beam systems.	30
2.5	Emission spectra of low- and medium-pressure mercury vapor lamps. Note that the low-pressure output is relevant for low-pressure and LPHO lamp systems	36
3.1	Power measurement per lamp vs lamp current setting	68
3.2	Headloss and water level data for bioassay exemplar. (a) System headloss as a function of flow and (b) system performance with varying water level and UVT headloss per bank vs flowrate.	69
3.3	The (a) MS2 and (b) T1 collimated beam results, respectively.	71
3.4	Comparison of predicted and measured log I values for the UV system from Manufacturer XYZ that was tested with a bioassay.	75
4.1	Schematic illustration of factors that affect performance in UV disinfection systems	82
4.2	Simulated particle trajectories and corresponding trajectory-specific UV doses, as developed by mapping a numerical simulation of the fluence rate field onto simulated particle trajectories.	84
4.3	Illustration of components of a CFD-I simulation for a single-lamp UV disinfection system	85
4.4	Photochemical transformation of *S* to yield *P*, as applied in Lagrangian actinometry	88
4.5	Measured variability on UV dose-response behavior of *E. coli* in a municipal wastewater effluent	91
5.1	(a) The effect of particle "shielding" on UV disinfection and (b) the effect of particles enmeshed/associated with microorganisms on UV disinfection	102

5.2	Particle size distribution of several biological treatment processes	105
5.3	Variation in UVT at Santa Rosa, California	107
5.4	(a) and (b) Collimated beam data for total coliform, enterococcus, and fecal coliform vs UV dose in unfiltered and filtered effluent	112
5.5	The effect of a prefiltration coagulation step on filtered effluent water quality	113
5.6	Anywhere WRRF secondary effluent collimated beam test results	122
6.1	Germicidal output as a function of lamp aging for a typical low-pressure high-output mercury lamp	135
7.1	Startup and commissioning activities	160

List of Tables

1.1	Recommended recreational water quality criteria	8
1.2	Summary of related references..	12
2.1	Ultraviolet dose requirements for 1- to 4-log credit disinfection of some pathogens and regulatory-important indicator organisms	34
3.1	Format and equivalency of the three key UV validation protocols commonly used for wastewater applications	53
3.2	Information that a bioassay of a UV system is capable or not capable of providing	55
3.3	Ultraviolet dose (fluence) to achieve 2-log (99%) reduction of select challenge microorganisms for UV disinfection systems	56
3.4	Description of the UV system from Manufacturer XYZ that was tested with a bioassay	66
3.5	Validation and operating range for use of the log I equation for the example UV disinfection system validation report	76
4.1	Summary of current guidance relative to UV disinfection design and validation	80
4.2	Estimated annual energy costs associated with reactor validation by biodosimetry (using coliphage T1 as the challenge organism) and those associated with validation by Lagrangian actinometry	90
5.1	Effect of treatment process on particle-associated bacteria	105
5.2	Ultraviolet-absorbing and light-scattering chemicals	108
5.3	Ultraviolet design and system configuration for Anywhere WRRF, U.S.A.	121
6.1	Ultraviolet disinfection system comparative capital costs	138
6.2	Typical design and guaranteed lives of significant UV components	139
6.3	Estimated annual costs at 62-ML/d (16-mgd) average daily flow	139
6.4	Summary of net present cost for example system	140
7.1	Summary of equipment procurement methods	159

8.1	Typical operational tasks	168
8.2	Typical alarm conditions for UV disinfection systems	176
8.3	Typical maintenance tasks	181
8.4	Troubleshooting checklist	186
9.1	Summary of UV system design case study examples	192
9.2	Summary of information on the Owen, Wisconsin, water resource recovery facility, representing an open-channel system with lamps oriented in a horizontal configuration	193
9.3	Summary of information on the Napolean, Ohio, water resource recovery facility, representing an open-channel system with lamps oriented in a horizontal configuration	194
9.4	Summary of information on the Manukau water resource recovery facility in Auckland, New Zealand, representing an open-channel system with lamps oriented in a horizontal configuration	195
9.5	Summary of information on the Penn Hills/Plum Creek water resource recovery facility, Pennsylvania, representing an open-channel system with lamps oriented in a horizontal configuration	196
9.6	Summary of information on the City of Jefferson, Missouri, Regional Water Reclamation Facility, representing an open-channel system with lamps oriented in a vertical configuration	197
9.7	Summary of information on the Auburn water resource recovery facility, representing an open-channel system with lamps oriented in an inclined configuration	198
9.8	Summary of information on the Rensselaer water resource recovery facility, representing an open-channel system with lamps oriented in an inclined configuration	199
9.9	Summary of information on the Macon, Missouri, water resource recovery facility, representing an open-channel system with microwave lamps oriented in a vertical configuration	200
9.10	Summary of information on Sieldlce Waterworks, Poland, representing a closed-vessel system	201
9.11	Summary of information on the Horizon, Texas, water resource recovery facility, representing a closed-vessel system	202
9.12	Summary of information on the Flat Creek water resource recovery facility, representing a closed-vessel system	203

Preface

The Water Environment Federation Disinfection and Public Health Committee has specifically identified a gap in the guidance that is available for UV disinfection systems for low dose applications, which includes disinfection of secondary and tertiary wastewater effluent discharges. This special publication provides information for engineers and wastewater utilities interested in using UV for disinfection and operators, with introductory information on UV disinfection. It also provides background information for regulatory agencies who review applications for approval of UV disinfection systems in water resource recovery facilities that are subject to discharge limits for bacteria.

This document, collaboratively developed by industry experts, includes information on UV technology as well as significant UV concepts such as bioassay validation and use of appropriate bioassay microorganisms. There are a number of alternative approaches that have been proposed and used for UV equipment sizing, including multiorganism bioassay techniques that provide the ability to design UV disinfection systems for the target pathogen or indicator rather than a default organism, and these concepts are explained. Additionally, considerations for design are presented, including an example of how to determine system sizing using a bioassay. Ultraviolet disinfection system redundancy, reactor layout and configuration, system procurement and construction, startup, and operations and maintenance are also described. The publication will provide information that is common to both large and small systems and will include case studies contributed by participating UV manufacturers and consulting engineers involved with design of low dose UV disinfection applications.

This publication was produced under the direction of Katherine (Kati) Y. Bell, Ph.D., P.E., BCEE, *Chair*.

Authors' and reviewers' efforts were supported by the following organizations:

AECOM, Burnaby, British Columbia, Canada
Alfa Laval, Inc., Richmond, Virginia
Aquionics, Inc., Erlanger, Kentucky
ARCADIS U.S., Inc., Indianapolis, Indiana
Black & Veatch Corporation, Kansas City, Missouri
Bratz Environmental Engineering, Boise, Idaho
Calgon Carbon Corporation, Markham, Ontario, Canada
Carollo Engineers, Sacramento, California, and Walnut Creek, California

Conestoga-Rovers & Associates, Shelby Township, Michigan
CDM Smith, Denver, Colorado; Nashville, Tennessee; and Tampa, Florida
CH2M HILL, Chantilly, Virginia; Santa Ana, California; Toronto, Ontario, Canada; and Vancouver, British Columbia, Canada
City of Atlanta, Atlanta, Georgia
City of Columbus, Columbus, Ohio
City of Toronto, Toronto, Ontario, Canada
Civil & Environmental Consultants, Inc., Charlotte, North Carolina
CSA Group, San Juan, Puerto Rico
DLZ Kentucky, Inc., Louisville, Kentucky
ENAQUA, Vista, California
Gray & Osborne, Inc., Seattle, Washington
Hazen and Sawyer, Raleigh, North Carolina
HDR, Mahwah, New Jersey, and Folsom, California
IDEXX Water, Westbrook, Maine
Johnson Controls, Inc., Westerville, Ohio
Johnson, Mirmiran & Thompson, Inc., Newark, Delaware
KCI Technologies, Inc., Fulton, Maryland
Leidos Engineering, LLC, Brookfield, Wisconsin
Lockwood, Andrews & Newman, Inc., Houston, Texas
Macon Water Authority, Macon, Georgia
McGill University, Montreal, Quebec, Canada
Purdue University, West Lafayette, Indiana
SEH, St. Paul, Minnesota
Stantec Consulting Services, Inc., Columbus, Ohio
Tel Aviv University, Israel
UltraTech Systems, Inc., Carmel, New York
Unitywater, Caboolture, Queensland, Australia
University of Colorado Boulder, Boulder, Colorado
University of Florida TREEO Center, Gainesville, Florida
University of Toledo, Toledo, Ohio
URS Corporation, Columbus, Ohio
U.S. Environmental Protection Agency, Washington, D.C.
Washington Suburban Sanitary Commission, Laurel, Maryland
XCG Consultants Ltd., Kitchener, Ontario, Canada
Xylem, Rye Brook, New York

1

Introduction

Robert Bastian and Katherine (Kati) Y. Bell, Ph.D., P.E., BCEE

1.0 PURPOSE	1	2.0 REGULATORY CONSIDERATIONS	9
1.1 Perspective	2	2.1 Regulatory Frameworks in the United States and Canada	9
1.2 History of Ultraviolet Disinfection	3		
1.3 Disinfection Criteria	4	2.2 Regulatory Drivers for Ultraviolet Disinfection	10
1.3.1 Indicator Bacteria	5		
1.3.2 Enumeration Methods for Indicator Bacteria	8	3.0 OTHER RELATED GUIDANCE	11
		4.0 ORGANIZATION OF THE PUBLICATION	13
		5.0 REFERENCES	13

1.0 PURPOSE

The purpose of this publication is to address considerations for UV disinfection systems designed to meet bacterial compliance for wastewater discharges to receiving waters and other wastewater disinfection applications where low UV doses are required to meet treatment objectives. This document does not address UV disinfection for drinking water or reuse applications where there is intimate human contact with the treated effluent that requires high UV doses to achieve treatment objectives (e.g., potable reuse or irrigation of food crops that may be consumed raw) or high-rate disinfection for treatment of wet weather flows that have not received full secondary biological treatment (e.g., primary effluent or blended flows). In the case of reuse, there are other reference documents available including the National Water Research Institute's and Water Research Foundation's (2003, 2012) *Ultraviolet Disinfection Guidelines for Drinking Water and Water Reuse.* For wet weather flows, UV disinfection may be incorporated as part of a holistic treatment approach that also includes processes that address the variable and sometimes high solids concentrations associated with this process. At this time, there is limited guidance specifically on the design and operations of UV disinfection for wet weather flows that have not received full secondary biological treatment, but the topic is addressed briefly in *Wet Weather Design and Operation in Water Resource Recovery Facilities* (WEF, 2014).

This publication provides information on UV disinfection for wastewater discharges that are regulated with criteria that have been established to protect human health and the environment. Thus, when disinfection standards are being met at their sources, public safety and water quality will be protected. It is also important to consider that there are factors that influence and potentially bias treatment efficacy; these factors include equipment design, operator training, equipment dependability, and operator attention. This is why it is necessary to provide guidance on UV disinfection of wastewater, which, after chlorine, is the most widely used method for wastewater disinfection (Leong et al., 2008). Thus, with respect to the aforementioned stated purpose, this document will be useful to a broad audience, including water resource recovery (WRRF) operators, engineers and designers, regulators, and the scientific community.

1.1 Perspective

Population growth and other increases on demands for water supply and water recreational uses significantly increase the opportunity for human exposure to wastewaters being discharged into the environment. Natural safeguards, such as dilution and distance or time before contact or use, have been reduced because of the larger volumes of wastewater being discharged and the number of discharge locations. Domestic wastewaters carry human pathogens excreted in fecal discharges of infected individuals, and even treated effluents can affect sources of domestic water supply, recreational waters, and shellfish growing areas. Many WRRFs have historically discharged their effluents to streams that are designated as recreational waters or are tributaries of larger waterbodies (streams or lakes) that are recreational waters or that are used as water supply sources by downstream communities. Potable water supplies are often extracted from these waterbodies, physically and chemically treated, and distributed to customers. While drinking water plants provide additional disinfection, the only protection recreational users receive is through adequate disinfection of wastewater effluents and, as such, disinfection is necessary to reduce potential transmission of infectious diseases when human contact is possible.

Disinfection, in the context of wastewater treatment as described in this publication, aims to reduce pathogen concentrations to levels where human health risks are acceptable. Thus, the objective of wastewater disinfection is not sterilization, which is inactivation of all microorganisms, rather a reduction in the concentrations of viable, pathogenic microorganisms that are responsible for the spread of illness/disease. Thus, to reduce the risks associated with fecal contamination, a number of disinfection methods are available that can be applied to wastewater effluents. Wastewater disinfection is rooted in the protection of human health and maintenance of a natural, healthy environment. Inactivation or destruction of pathogenic microorganisms at municipal WRRFs can reduce the dissemination of pathogens to the environment and break the potential cycles of pathogen-associated infections.

Chlorination became the standard process for disinfecting treated wastewater effluents and was key to the great public health successes of the 20th century. However, with awareness of the environmental effects associated with disinfection practices in the 1960s and 1970s, regulators determined that the ongoing need to effectively destroy pathogenic microorganisms must be balanced against the effects of a disinfected wastewater effluent on the biota in receiving water and the creation of byproducts that had serious public health consequences (Whitby and Scheible, 2004). The deleterious effects of halogens on the environment, along with the long-term effects of halogenated hydrocarbons, have been well documented. In the 1970s, this prompted governments in Canada and the United States (Environment Canada, 1978; U.S. EPA, 1976) to introduce rules regarding halogenated disinfection byproducts that have reduced the use of chlorine as a disinfectant for wastewater (Riordan, 1979). More restrictive limits began to be placed on chlorine residuals, often requiring dechlorination before discharge. The investigation and implementation of alternative disinfection methods, such as UV and ozone, was encouraged, prompting a considerable amount of research and demonstration efforts with alternative disinfectants (Whitby and Scheible, 2004). Thus, with advances in UV disinfection technologies for wastewater applications, not only have the economics of UV disinfection become more favorable, but the operation and maintenance of UV disinfection systems may also be much safer and simpler than many alternative technologies.

1.2   History of Ultraviolet Disinfection

Downes and Blunt (1877) made the first recorded discovery of the bactericidal effects of sunlight in England. Engineered UV systems were made possible with the invention of the mercury vapor arc lamp by Peter Cooper Hewitt in 1901, coupled with a quartz sleeve in 1906, which led to the production of the first commercial UV lamps. The first recorded use of UV light for disinfection of water was in Marseilles, France, in 1910 using a Westinghouse Cooper Hewitt mercury lamp in fused quartz. Drinking water disinfection was initially the focus of UV applications where, in 1916, UV light was used for the disinfection of water on ships. Numerous UV disinfection facilities in the United States were installed and operated between 1916 to 1939 in places such as New York, Kentucky, Ohio, and Kansas (Baker, 1948). These facilities were largely abandoned because of operational costs, problems with reliable electrical supplies, and the emergence of chlorine as an effective technology for pathogens of concern at that time. While UV disinfection did not find a practical reemergence in the United States until the 1970s when it was explored for wastewater disinfection, the scientific understanding of UV-based photochemistry and photobiology underwent intense growth. Deep understanding of the inactivation efficacy of UV light, including its fundamental effects on nucleic acid damage, were generated in the 1920s through the 1950s, providing the basis of much of our understanding of how UV disinfection works.

Whitby and Scheible (2004) published a detailed history of UV disinfection focusing on the practice of UV disinfection of wastewater from the late 1970s to the present. Advances in ballast and electronics technology in the 1970s and 1980s, along with the desire to find disinfection alternatives to chlorine, which was of concern in environmental discharges because of the formation of chlorinated disinfection byproducts, helped the resurgence of UV light treatment as a viable technology for disinfection. Ultraviolet disinfection has since grown into a significant commercial industry and is the subject of much academic and industrial research. Whitby and Scheible note two significant milestones in the acceptance of UV disinfection of wastewater: successful demonstration of a full-scale innovative UV system in 1978 at the Northwest Bergen Wastewater Treatment Plant in Waldwick, New Jersey (Scheible and Bassell, 1981), and introduction of a modular UV system for wastewater in 1982 in Tillsonburg, Ontario, Canada, that fit within a gravity-fed, open channel with lamps parallel to the flow (Whitby et al., 1984). Since the early 1980s, UV light has developed a large market share in wastewater disinfection applications, where it has been shown to be competitive with chlorination in terms of disinfection efficacy and economics (Darby et al., 1995). By 1985, there was a jump in the application of UV light for smaller WRRFs as UV technology was deemed proven and reliable. Use of UV disinfection for wastewater makes sense for a variety of reasons. It has a small footprint, there is no need for a large contact basin, it eliminates the need for dechlorination before discharge into a natural waterbody, and it is easy to use.

1.3 Disinfection Criteria

Public health agencies worldwide have long understood the relationship between fecal contamination in surface waters and the associated human health risks. Because of the difficulties in identifying the specific origin of illnesses associated with fecal contamination, as early as the 1960s, the U.S. Public Health Service (USPHS) recommended using fecal coliform bacteria as an indicator for human health risks associated with primary contact. This recommendation was based on studies that reported a detectable health effect when total coliform densities exceeded about 2300 colony-forming units (cfu)/100 mL (Stevenson, 1953).

Whereas these correlations between fecal coliform bacteria and waterborne illnesses have been documented, interestingly, it is also known that most strains of fecal indicator bacteria (i.e., those that are used for the purposes of monitoring) are not pathogenic. It is the presence of other pathogenic organisms (such as viruses, pathogenic bacteria, and protozoa) that cause these illnesses. Thus, while many species of fecal coliform bacteria are not pathogenic, these microorganisms demonstrate characteristics that make them good indicators of fecal contamination (i.e., often of fecal origin and simple methods of detection) and, therefore, indirectly indicate the potential presence of fecal pathogens capable of causing illnesses. As such, the fecal indicator bacteria are "indicators" of the potential for human infectious diseases. Scientists recognize that the use of an indicator is not a perfect method for detecting the presence of all of the numerous pathogens that cause illnesses associated with human exposure to surface waters where wastewater is discharged. However, use of these indicators is supported by epidemiological studies on human health relationships and this approach overcomes issues associated with pathogen-specific enumeration methods for environmental waters (U.S. EPA, 2012). Furthermore, indicator organisms have often served as a criterion that is the basis of a regulatory framework for wastewater disinfection.

1.3.1 Indicator Bacteria

The rationale for using indicator organisms as the basis for microbiological criteria is that, with the epidemiological knowledge currently available, it is difficult to assess the specific risk to health presented by any particular level of pathogens in water because this risk will depend on the infectivity and invasiveness of the pathogen and the innate and acquired immunity of individuals contacting the water. In addition, only certain waterborne pathogens can be detected reliably and easily in water, and some cannot be detected at all (WHO, 1996). Thus, the best indicators of fecal contamination will be those that are universally present in large numbers in the feces of humans and warm-blooded animals, that are readily detected by simple methods, do not grow in natural waters, and that persist in water and can be removed by wastewater treatment similar to waterborne pathogens.

In 1968, the National Technical Advisory Committee (NTAC) translated the previously established total coliform level of 2300 per 100 mL (Stevenson, 1953) to 400 fecal coliforms per 100 mL based on a ratio of total to fecal coliform, and then halved that number to 200 fecal coliforms per 100 mL (U.S. EPA, 2012). The NTAC criteria for recreational waters were then recommended again by the U.S. Environmental Protection Agency (U.S. EPA) in 1976, even though the criteria had been criticized for a number of issues related to the design of the USPHS studies and the limited amount of epidemiological data and data quality. The 1976 U.S. EPA criterion for bacteria in primary recreational waters required that fecal coliform content not exceed a geometric mean of 200 organisms per 100 mL and that no more than 10% of the total number of samples, taken during any 30-day period, exceeded 400 fecal coliforms per 100 mL (U.S. EPA, 1976). By 1986, as more data became available, U.S. EPA recommended that *Escherichia coli* and enterococci be used for assessing microbiological water quality in recreational waters because concentrations of these organisms are more strongly correlated with swimming-associated gastroenteritis rates (U.S. EPA, 1986). It should be noted that many states only apply the criteria seasonally to protect human health during the season in which human contact would occur.

In the United States, many states questioned whether they should adopt the 1986 recommendations for *E. coli* or enterococci for setting water quality standards, and some state regulators asked why it was necessary to change their programs if the estimation of disease risk to swimmers had not significantly improved. Because of new studies and data, U.S. EPA took the position that *E. coli* and enterococci were better indicators of public health risk in recreational waters than fecal coliforms. Results from epidemiological evidence firmly linked *E. coli* and enterococci levels to swimming-related illness (Cabelli, 1983; Dufour, 1984). When developing criteria based on *E. coli* and enterococci, U.S. EPA did not propose criteria that were more stringent than the 200 fecal coliforms per 100 mL value first recommended in 1968. Instead, they represented the disease risk estimated for swimmers at freshwater and marine beaches with exposures to the maximum fecal coliform limit. The 1986 criteria values were calculated to represent the ambient condition of the waterbody necessary to protect the designated use of primary contact recreation. These values were selected to carry forward the same level of water quality associated with U.S. EPA's previous criteria recommendations to protect the primary contact recreation use as that for fecal coliforms (U.S. EPA, 1976). The 1986 criteria also carried a single sample maximum (SSM) component, which was computed using the geometric mean values and corresponding observed variances in the fecal indicator bacteria obtained from water quality measurements taken during the epidemiological studies from the late 1970s and early 1980s. Four different SSM values (recommended to be used with different recreational use intensities) were provided and corresponded to different percentiles of the water quality distribution around the geometric mean. The 1986 criteria values were based on different water quality values and associated illness rates for marine and fresh waters because the marine and freshwater epidemiological studies reported different geometric mean values for the indicator bacteria associated with the water quality corresponding to U.S. EPA's fecal coliform criteria recommendations.

For decades, epidemiological studies have been used to evaluate how fecal indicator bacteria concentrations are associated with health effects of primary contact recreation on a quantitative basis. The aforementioned 1986 criteria recommendations are supported by epidemiological studies conducted by U.S. EPA in the late 1970s and early 1980s. In those studies, enterococci and *E. coli* exhibited the strongest correlations to swimming-associated gastroenteritis. Both of these indicators continue to be used in epidemiological studies conducted throughout the world, including in the European Union and Canada (EP/CEU, 2006). The World Health Organization (WHO) recommends the use of enterococci as water quality indicators for recreational waters (WHO, 2003). Meta analyses and systematic reviews of epidemiological studies conducted worldwide indicate that these indicators generally provided substantial improvements over the indicators that were favored previously, such as total and fecal coliforms (Prüss, 1998; Wade et al., 2003; Zmirou et al., 2003). It should be noted that total and fecal coliforms, as indicators, also include other bacteria such as *Klebsiella* that are not necessarily fecal in origin; *Klebsiella* are commonly associated with textile and pulp and paper mill wastes. Thus, when U.S. EPA most recently updated its recreational water quality criteria (RWQC) in 2012, enterococci and *E. coli* were again recommended for fresh water, and enterococci as the indicator to be measured in both marine and fresh water. The most recent U.S. EPA (2012) RWQC offers two sets of recommended numeric concentration thresholds, either of which would be protective for primary contact in recreational waters.

The criteria recommended in the 2012 U.S. EPA RWQC (Table 1.1) would protect the public from exposure to harmful levels of pathogens; the illness rates that U.S. EPA recommended are based on the National Epidemiological and Environmental Assessment of Recreational Water definition of gastrointestinal illness, which is limited to illnesses that exhibit a fever. This study allowed U.S. EPA to provide better estimates of risk based on the new data. These recommendations have been issued as guidance to states, territories, and authorized tribes for use in developing water quality standards to protect swimmers from exposure to water that contains

TABLE 1.1 Recommended recreational water quality criteria (U.S. EPA, 2012).

Criteria elements Indicator	Estimated illness rate: 36 per 1000 primary contact recreators		Estimated illness rate: 32 per 1000 primary contact recreators	
	Magnitude		Magnitude	
	Geometric mean (cfu/100 mL)	STV (cfu/100 mL)	Geometric mean (cfu/100 mL)	STV (cfu/100 mL)
Enterococci—marine and fresh	35	130	30	110
E. coli—fresh	126	410	100	320

Duration and frequency: The waterbody geometric mean should not be greater than the selected geometric mean magnitude in any 30-day interval. There should not be greater than a 10% excursion frequency of the selected STV magnitude in the same 30-day interval.

organisms that indicate the presence of fecal contamination. It should be noted that the SSM has been dropped in the 2012 U.S. EPA RWQC in favor of a statistical threshold value (STV); new criteria are comprised of both a geometric mean and the STV. For a set of criteria values, U.S. EPA computed the STV based on water quality distribution observed during U.S. EPA's epidemiological studies. The STV approximates the 90th percentile of the water quality distribution and is intended to be a value that should not be exceeded by more than 10% of the samples used to calculate the geometric mean. Because densities of fecal indicator bacteria are highly variable in ambient waters, distributional estimates are more robust than single-point estimates (U.S. EPA, 2012).

1.3.2 Enumeration Methods for Indicator Bacteria

Indicators of fecal contamination are detected and enumerated using a variety of methods. Fecal indicator bacteria, which, as described previously, are used as the basis of monitoring in discharge permits, can be enumerated using various analytical methods including those in which the organisms are grown (cultured) and those in which their DNA is extracted from an environmental sample, amplified, and quantified (using quantitative polymerase chain reaction [qPCR]). These different enumeration methods result in method-specific units and values. One culture-based method, membrane filtration, yields a number of colonies that arise from bacteria captured on the membrane filter per volume of water filtered; one colony can be produced from one or several cells (clumped cells in the environmental sample) and results are expressed in units such as colony-forming units per volume (e.g., cfu/100 mL). Other culture-based measurements, based on use of a defined substrate method, produce a most probable number (MPN) per volume; bacterial densities' MPNs are based on the combination of positive and negative test tube results that can be read from an MPN table (U.S. EPA, 1978) such as the IDEXX Quanti-Tray®, which provides bacteria results as an MPN. Results from qPCR analyses are reported in units that are calculated based on the target DNA sequences from test samples relative to those in calibrator samples that contain a known quantity of target organisms (Haugland et al., 2005; Wade et al., 2010).

There are challenges with respect to interpreting results from each of these enumeration techniques because each analytical technique can vary depending on a number of test factors. For example, culture-based methods depend on the metabolic state (i.e., viability and activity) of the target organisms for effective enumeration. In contrast, qPCR-based approaches detect specific sequences of DNA that have been extracted from a water sample, and the results contain sequences from both viable and nonviable forms of the targeted indicator, which is highly problematic for monitoring performance of UV disinfection and may only inactivate microorganisms rather than "kill" the organisms; it is important to note that operationally, UV-inactivated organisms can no longer divide and multiply, and thus are not infectious or pathogenic.

2.0 REGULATORY CONSIDERATIONS

Globally, the driver for disinfection is the protection of human health from recreational contact with potentially pathogenic organisms in waterbodies and/or to protect water supplies from being contaminated with pathogenic microorganisms. In the United States and Canada, wastewater discharges are regulated through a framework that is implemented at a state or province level. Although the regulatory approach varies globally, in general there are criteria, such as those set by WHO (2003), that provide guidance on minimum treatment requirements for wastewater and that are generally consistent with the U.S. EPA criteria.

2.1 Regulatory Frameworks in the United States and Canada

In the United States, the Clean Water Act (CWA) set a national goal "to restore and maintain the chemical, physical, and biological integrity of the Nation's waters", with an interim goal that all waters be "fishable and swimmable" where possible. The CWA specifies that all point-source discharges into the nation's waters are unlawful unless authorized by a National Pollution Discharge Elimination System permit and sets baseline, technology-based controls for municipalities and industries. It requires all dischargers to meet additional, stricter pollutant controls where needed to meet water quality targets and requires federal approval of these standards. The CWA also provides the authority for U.S. EPA to establish specific criteria for protecting the public from exposure to harmful levels of pathogens while participating in water-contact activities such as swimming, wading, and surfing in waters designated for such uses. Thus, in the United States, it is recreational water quality uses that drive permit limits for disinfection.

In Canada, the Canadian Environmental Protection Act (CEPA) (Environment Canada, 1999) is the primary element of the legislative framework for protecting the Canadian environment and human health. A key aspect of CEPA 1999 is the prevention and management of risks posed by toxic and other harmful substances, including microorganisms. The Minister of the Environment is accountable to the Canadian Parliament for the administration of all of CEPA 1999. Both the Minister of the Environment and the Minister of Health jointly administer the task of assessing and managing the risks associated with toxic substances. Efforts taken under CEPA 1999 are complemented by actions taken under other federal acts administered by the minister of the environment. The Fisheries Act, administered by the Minister of the Environment on behalf of the Minister of Fisheries and Oceans, includes provisions to prevent pollution of waters inhabited by fish. Through the Canada Water Act, water resources and their environmental quality are also managed.

2.2 Regulatory Drivers for Ultraviolet Disinfection

In 2014, U.S. EPA issued an updated draft of its national recommended water quality criteria for human health for 94 chemical pollutants to reflect the latest scientific information and U.S. EPA policies; the update included new fish consumption rates and limits for disinfection byproducts that are more restrictive than the previously published criteria. Once finalized, U.S. EPA water quality criteria provide recommendations to states and tribes authorized to establish water quality standards under the CWA. In addition to disinfection byproducts, individual states are faced with increasingly stringent nutrient limits that have a direct effect on chlorine-based disinfection. Similar activities are requiring WRRFs in Canada to review their current practices of chlorine disinfection.

In 2006, Environment Canada began work on a national strategy, under direction of the Canadian Council of Ministers of the Environment (CCME), to provide additional requirements for management of wastewater effluents. In 2007, CCME released a "Draft Canada-wide Strategy for the Management of Municipal Wastewater Effluent". At the same time, Environment Canada published a "Proposed Regulatory Framework for Wastewater" to explain how the Canada-wide strategy would be implemented. A draft regulation based on the CCME strategy was released for comment in 2010 and, on July 28, 2012, the Wastewater Systems Effluent Regulations (WSER) were published in the *Canada Gazette* (Environment Canada, 2012). The regulations fall under the Fisheries Act, which prohibits unauthorized deterioration, disruption, and destruction of fish habitat. The WSER represents the first national regulations in Canada that specifically address municipal WRRF effluents that are regulated under provincial authorities. In the new Canadian WSER, these parameters include un-ionized ammonia, acute lethality testing, and total residual chlorine (TRC). As a result, facilities that have historically used chlorination may be faced with highly variable effluent ammonia and nitrites when solids retention time is increased to improve treatment for ammonia. A detailed discussion of the effects of these ions on chlorination and the disinfection process control using chlorine may be found in *White's Handbook of Chlorination and Alternative Disinfectants* (Black & Veatch, 2010).

Thus, with respect to WRRFs, emerging criteria and new regulations may impose limits for a number of parameters in final effluent that affect disinfection process control for chlorination. These challenges, coupled with the need to add dechlorination to meet toxicity or TRC limits, are drivers for many facilities that are looking to solve disinfection challenges associated with chlorination by converting to UV disinfection.

3.0 OTHER RELATED GUIDANCE

There are other uses of UV disinfection for bacterial inactivation in wastewater applications, including stormwater, wet weather flows, and noncontact reuse applications that may benefit from information in this guidance document. However, it is not the specific intent of this publication to address these applications and, therefore, it is recommended that appropriate state regulatory agencies be consulted where appropriate. The information presented in this publication aims to define key design considerations and leverage information and knowledge that have already been reported. While many of these industry references are specific to other applications (e.g., reuse or drinking water), some of the fundamentals and other information presented within those publications provide relevant background for development of this publication. Table 1.2 summarizes some of these references, along with their applicability; although the table does not provide a complete list of available literature, it can be used as a starting point for the reader to become educated on the broader areas of UV disinfection for wastewater and other applications.

TABLE 1.2 Summary of related references.

Application	References
General wastewater disinfection guidance, including various disinfection methods	*Design Manual: Municipal Wastewater Disinfection* (U.S. EPA, 1986) *Disinfection of Wastewater Effluent—Comparison of Alternative Technologies* (Leong et al., 2008) *White's Handbook of Chlorination and Alternative Disinfectants* (Black & Veatch, 2010)
General information on UV fundamentals and equipment components	*UV Disinfection Knowledge Base* (WRF, 2012)
Ultraviolet disinfection guidance for drinking water and reuse applications	*Ultraviolet Disinfection Guidance Manual for the Final Long Term 2 Enhanced Surface Water Treatment Rule* (U.S. EPA, 2006) *Ultraviolet Disinfection Guidelines for Drinking Water and Water Reuse* (NWRI and WRF, 2012) *Plants for the Disinfection of Water Using Ultraviolet Radiation—Requirements and Testing—Part 1: Low Pressure Mercury Lamp Plants* (ÖNORM, 2001) *Plants for the Disinfection of Water Using Ultraviolet Radiation—Requirements and Testing—Part 2: Medium Pressure Mercury Lamp Plants* (ÖNORM, 2003) *UV Disinfection Devices for Drinking Water Supply—Requirements and Testing* (DVGW, 2006)
Design and operations considerations for wastewater UV disinfection systems	*Operation of Municipal Wastewater Treatment Plants* (WEF, 2007) *Design of Municipal Wastewater Treatment Plants* (WEF et al., 2009) *The Effect of Upstream Treatment Processes on UV Disinfection Performance* (Darby, 1999)
Guidance for validation of UV disinfection equipment systems	*Uniform Protocol for Wastewater UV Validation Applications* (Whitby et al., 2011)

4.0 ORGANIZATION OF THE PUBLICATION

The chapters of this guidance publication are organized such that the reader develops an understanding of the theory of UV disinfection before reading about equipment systems and systems validation. With this background, the reader will have a better grasp of the content of the chapters that provide details on engineering design and construction of these systems. Information on UV process concepts are summarized in Chapter 2; low-dose bioassay methods and a bioassay exemplar are presented in Chapter 3; and Chapter 4 includes a discussion on innovations and advances in UV disinfection design approaches. Process design considerations and an example of UV system sizing are provided in Chapter 5; Chapter 6 includes a description of equipment selection and facilities design considerations; and Chapter 7 includes a discussion of procurement, construction, installation, and performance testing. Chapter 8 provides an overview of operational considerations for wastewater UV disinfection systems and Chapter 9 provides several case study examples that demonstrate information presented in previous chapters.

5.0 REFERENCES

Baker, M. N. (1948) *The Quest for Pure Water: The History of Water Purification from the Earliest Records to the Twentieth Century*; American Water Works Association: Denver, Colorado.

Black & Veatch (2010) *White's Handbook of Chlorination and Alternative Disinfectants*, 5th ed.; Wiley & Sons: Hoboken, New Jersey.

Cabelli, V. J. (1983) *Health Effects Criteria for Marine Recreational Waters*; EPA-600/1-80-031; U.S. Environmental Protection Agency: Washington, D.C.

Council of the European Communities (2006) Directive 2006/7/EC of The European Parliament and of The Council of 15th February 2006 concerning the management of bathing water quality and repealing Directive 76/160/EEC; *Off. J. Eur. Communities*, **L64**, 37–51.

Darby, J.; Heath, M.; Jacangelo, J.; Loge, F.; Swaim, P.; Tchobanoglous, G. (1995) *Comparison of UV Irradiation to Chlorination: Guidance for Achieving Optimal UV Performance*; Water Environment Research Foundation: Alexandria, Virginia.

Darby, J. (1999) *The Effect of Upstream Treatment Processes on UV Disinfection Performance*; Water Environment Research Foundation: Alexandria, Virginia.

Deutscher Verein des Gas und Wasserfaches (2006) *UV Disinfection Devices for Drinking Water Supply—Requirements and Testing;* DVGW W294-1, -2, and -3; Deutscher Verein des Gas und Wasserfaches: Bonn, Germany.

Dufour, A. P. (1984) *Health Effects Criteria for Fresh Recreational Waters;* EPA-600/1-84-004; U.S. Environmental Protection Agency: Washington, D.C.

Downes, A.; Blunt, T. P. (1877) Researches on the Effect of Light Upon Bacteria and Other Organisms. *Proc. Royal Soc. London,* **26**, 488–500.

Environment Canada (1999) Canadian Environmental Protection Act. http://www.ec.gc.ca/lcpe-cepa/default.asp?lang=En&n=CC0DE5E2-1&toc=hide (accessed Jan 2015).

Environment Canada (1978) *Wastewater Disinfection in Canada;* EPS 3-WP-78-4; Environmental Protection Service: Gatineu, Quebec, Canada.

Environment Canada (2012) Wastewater Systems Effluent Regulations. http://laws-lois.justice.gc.ca/eng/regulations/SOR-2012-139/FullText.html (accessed Jan 2015).

Haugland, R. A.; Siefring, S. C.; Wymer, L. J.; Brenner, K. P.; Durfour, A. (2005) Comparison of Enterococcus Measurements in Fresh Water at Two Recreational Beaches by Quantitative Polymerase Chain Reaction and Membrane Filter Culture Analysis. *Water Res.,* **39**, 559–568.

Leong, L. Y. C.; Kuo, J.; Tang, C.-.C. (2008) *Disinfection of Wastewater Effluent—Comparison of Alternative Technologies;* WERF Report No. 04-HHE-4; Water Environment Research Foundation: Alexandria, Virginia.

National Water Research Institute; Water Research Foundation (2003) *Ultraviolet Disinfection Guidelines for Drinking Water and Water Reuse,* 2nd ed.; National Water Research Institute: Fountain Valley, California.

National Water Research Institute; Water Research Foundation (2012) *Ultraviolet Disinfection Guidelines for Drinking Water and Water Reuse,* 3rd ed.; National Water Research Institute: Fountain Valley, California.

ÖNORM (2001) *Plants for the Disinfection of Water Using Ultraviolet Radiation—Requirements and Testing—Part 1: Low Pressure Mercury Lamp Plants;* ÖNORM M 5873-1; Osterreichisches Normungsinstitut: Vienna, Austria.

ÖNORM (2003) *Plants for the Disinfection of Water Using Ultraviolet Radiation—Requirements and Testing—Part 2: Medium Pressure Mercury Lamp Plants;* ÖNORM M 5873-2; Osterreichisches Normungsinstitut: Vienna, Austria.

Prüss, A. (1998) Review of Epidemiological Studies on Health Effects from Exposure to Recreational Water. *Int. J. Epidemiology,* **27** (1), 1–9.

Riordan, C. (1979) *Perspectives on Wastewater Disinfection—A View from Headquarters*. Progress in Wastewater Disinfection Technology; EPA-600/9-79-018; U.S. Environmental Protection Agency: Cincinnati, Ohio.

Scheible, O. K.; Bassell, C. D. (1981) *Ultraviolet Disinfection of a Secondary Wastewater Treatment Plant Effluent*; EPA-600/S2-B1-152; U.S. Environmental Protection Agency: Cincinnati, Ohio.

Stevenson, A. H. (1953) Studies of Bathing Water Quality and Health. *Am. J. Public Health*, **43**, 429–538.

U.S. Environmental Protection Agency (1986) *Design Manual: Municipal Wastewater Disinfection*; EPA-625/1-86-021; U.S. Environmental Protection Agency, Office of Research and Development, Water Engineering Research Laboratory Center for Environmental Research Information: Cincinnati Ohio.

U.S. Environmental Protection Agency (1976) *Disinfection of Wastewater—Task Force Report*; EPA-430/9-75-013; U.S. Environmental Protection Agency: Washington, D.C.

U.S. Environmental Protection Agency (1978) *Microbiological Methods for Monitoring the Environment: Water and Wastes*; EPA-600/8-78-017; U.S. Environmental Protection Agency, Office of Research and Development: Cincinnati, Ohio.

U.S. Environmental Protection Agency (2012) *Recreational Water Quality Criteria*; EPA-820/F-12-058; U.S. Environmental Protection Agency: Washington, D.C.

U.S. Environmental Protection Agency (2006) *Ultraviolet Disinfection Guidance Manual for the Final Long Term 2 Enhanced Surface Water Treatment Rule*; EPA-815/R-06-007; U.S. Environmental Protection Agency: Washington, D.C.

Wade, T. J.; Sams, E.; Brenner, K. P.; Haugland, R.; Chern, E.; Beach, M.; Wymer, L.; Rankin, C. C.; Love, D.; Li, Q.; Noble, R.; Dufour, A. P. (2010) Rapidly Measured Indicators of Recreational Water Quality and Swimming-Associated Illness at Marine Beaches: A Prospective Cohort Study. *Environ. Health*, **9**, 66.

Wade, T. J.; Pai, N.; Eisenberg, J. N. S.; Colford, J. M., Jr. (2003) Do U.S. Environmental Protection Agency Water Quality Guidelines for Recreational Waters Prevent Gastrointestinal Illness? A Systematic Review and Meta-Analysis. *Environ. Health Perspectives*, **111** (8), 1102–1109.

Water Environment Federation; American Society of Civil Engineers; Environmental and Water Resources Institute (2009) *Design of Municipal Wastewater Treatment Plants*, 5th ed.; WEF Manual of Practice No. 8;

ASCE Manual and Report on Engineering Practice No. 76; McGraw-Hill: New York.

Water Environment Federation (2007) *Operation of Municipal Wastewater Treatment Plants;* WEF Manual of Practice No. 11; Water Environment Federation: Alexandria, Virginia.

Water Environment Federation (2014) *Wet Weather Design and Operation in Water Resource Recovery Facilities;* Water Environment Federation: Alexandria, Virginia.

Water Research Foundation (2012) *UV Disinfection Knowledge Base;* Water Research Foundation: Denver, Colorado.

Whitby, G. E.; Lawal, O.; Ropic, P.; Shmia, S.; Ferran, B.; Dussert, B. (2011) Uniform Protocol for Wastewater UV Validation Applications. *IUVA News,* **13** (2), 26–33.

Whitby, G. E.; Palmateer, G.; Cook, W. G.; Maarschalkerweerd, J.; Huber, D.; Flood, K. (1984) Ultraviolet Disinfection of Secondary Effluent. *J. Water Pollut. Control. Fed.,* **56** (7), 844–850.

Whitby, G. E.; Scheible, O. K. (2004) The History of UV and Wastewater. *IUVA News,* September 2004, 15–26.

World Health Organization (2003) Guidelines for Safe Recreational Water Environments; Volume 1: Coastal and Fresh Waters. http://www.who.int/water_sanitation_health/bathing/srwe1/en/ (accessed Jan 2015).

World Health Organization (1996) Guidelines for Drinking-Water Quality. Volume 2: Health Criteria and Other Supporting Information, 2nd ed.; World Health Organization: Geneva, Switzerland.

Zmirou, N.; Pena, L.; Ledrans, M.; Leterte, A. (2003) Risks Associated with the Microbiological Quality of Bodies of Fresh and Marine Water Used for Recreational Purposes: Summary Estimates Based on Published Epidemiological Studies. *Arch. Environ. Health,* **58** (11), 703–711.

2

Ultraviolet Disinfection Process Concepts and Equipment Systems

Karl G. Linden, Ph.D., and Hadas Mamane, Ph.D.

1.0 PRINCIPLES OF ULTRAVIOLET DISINFECTION	18	
1.1 Electromagnetic Spectrum	18	
1.2 Properties of Ultraviolet Light	19	
1.2.1 Properties of a Photon	19	
1.2.2 Ultraviolet Absorbance and Transmittance	20	
1.2.3 Laws of Photochemistry	21	
1.2.4 Ultraviolet Scattering	22	
1.3 Germicidal Action of Ultraviolet Light	22	
1.3.1 Nucleic Acid Damage	23	
1.3.2 Deoxyribonucleic Acid and Protein Absorbance	24	
1.3.3 Ultraviolet Action Spectrum	25	
1.3.3.1 Means to Obtain an Action Spectrum	25	
1.3.3.2 Action Spectrum by Organism	26	
1.3.3.3 Can the Deoxyribonucleic Acid Absorbance Spectrum Represent the Action Spectrum?	26	
1.4 Microbial Repair and Regrowth	27	
1.4.1 Photo-Reactivation	28	
1.4.2 Dark Repair	29	
1.4.3 Regrowth	29	
2.0 ULTRAVIOLET DOSE RESPONSES OF PATHOGENS AND SURROGATES	30	
2.1 Collimated Beam Testing	30	
2.1.1 Types of Collimated Beams	30	
2.1.2 Ultraviolet Dose Measurements and Calculation	30	
2.1.3 Factors Affecting Ultraviolet Dose Calculation	32	
2.2 Data on Microbe Dose Response	33	
2.2.1 Bacteria	34	
2.2.2 Viruses	34	
2.2.3 Protozoa	35	
3.0 ULTRAVIOLET LAMP TECHNOLOGIES	35	
3.1 Low-Pressure Ultraviolet Lamps	36	
3.2 Low-Pressure High-Output Ultraviolet Lamps	37	

3.3	Medium-Pressure Ultraviolet Lamps	37	3.4.3	Pulsed Ultraviolet Lamps	39
3.4	Alternative Lamp Technologies	37	3.4.4	Excimer Lamps	40
	3.4.1 Ultraviolet Light-Emitting Diodes	38	4.0	REFERENCES	40
	3.4.2 Microwave Ultraviolet Radiation	38			

1.0 PRINCIPLES OF ULTRAVIOLET DISINFECTION

1.1 Electromagnetic Spectrum

The sun's electromagnetic radiation contains a wide range of wavelengths. Ultraviolet radiation is the region of the electromagnetic spectrum that lies between X-ray and visible light, at 100 to 400 nm (U.S. EPA, 2006). The UV spectrum of the sunlight is divided into four regions: vacuum UV (VUV) (<200 nm), UV C (UVC) (200 to 280 nm), UV B (UVB) (280 to 315 nm), and UV A (UVA) (315 to 400 nm) (Figure 2.1). The UV region below 200 nm is mostly absorbed by air and, therefore, it is necessary to use a vacuum to work at these wavelengths (hence the term, *vacuum UV*) (Meulemans, 1987). However, all of UVC light, and most of UVB light (above 95%), are absorbed by the ozone layer that function as a filter before reaching the earth's surface, providing complete attenuation under 300 nm.

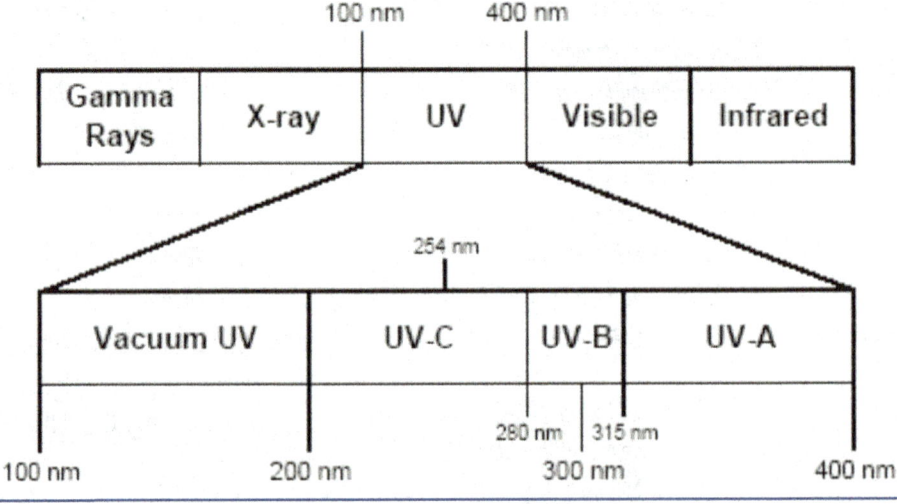

FIGURE 2.1 Ultraviolet light in the electromagnetic spectrum (U.S. EPA, 2006).

Thus, nearly all of the terrestrial sunlight UV radiation reaching the earth's surface comprises UVA (95%) and UVB (5%).

Infrared radiation (wavelength above 700 nm) and UV radiation (below 400 nm) are best known for their capability to destroy or inactivate cells by thermo-chemical and photochemical reactions, respectively. Infrared and UV radiation cause structural and/or functional changes of nucleic acids and proteins and alteration of other cellular components, ultimately damaging the cells in a reversible or irreversible manner (Belliveau et al., 1992; Berney et al., 2006; Berney et al., 2007; Patterson and Gillespie, 1972; Pfeifer et al., 2005; Yoon et al., 2000). Ultraviolet A generates reactive oxygen species and free radicals in cells (Foyer et al., 1994; Paul and Gwynn-Jones, 2003; Sinha and Häder, 2002). These molecules possess high destructive potential because of their tendency to react with many of the cell's components including lipids, proteins, and nucleic acids. Cell membrane destruction was also reported (Sinha and Häder, 2002). Nevertheless, most of the damaging effects of UV irradiation to microorganisms are attributed to UVB and UVC absorption by cellular DNA.

1.2 Properties of Ultraviolet Light

Radiation originates from the emission of matter, and its transport does not require the presence of matter. One theory views radiation as a propagation of particles termed *photons*; alternatively, radiation can be viewed as a propagation of electromagnetic waves, with properties of frequency and wavelength. According to the quantum theory, light is also quantized. The absorption or emission of light occurs by the transfer of energy as photons (Wardle, 2009).

1.2.1 Properties of a Photon

Radiation travels through space in discrete units called *quanta* or *photons* of energy. The energy of each photon is related to the wavelength of the radiation, as described by Planck's Law. Photons have both wave-like and particle-like properties and each photon has a specific energy. The energy E, quantum of radiation, represents the difference between the final and initial energy states and is proportional to the frequency of radiation. The tie between the wave and quantum theories is reflected by the fact that a quantum of light has its energy calculated using the wavelength of light (Calvert and Pitts, 1966). The energy of a photon is proportional to its frequency and is inversely proportional to its wavelength, as follows:

$$E = h \cdot v = \cdot \frac{c}{\lambda} \qquad (2.1)$$

Where
> E = the photon energy (J),
> h = Planck's Constant, 6.63×10^{-34} (J s),
> v = the frequency of light (s),
> c = the speed of light, 2.998×10^8 m/sec, and
> λ = the wavelength of light, m.

The unit Einstein is designated to represent a mole of photons.

1.2.2 Ultraviolet Absorbance and Transmittance

In direct photolysis (photodegradation), a molecule absorbs a photon, resulting in transition of the molecule from a ground electronic state to an excited electronic state. For photon absorption to be possible, the energy of the photon (inversely proportional to its wavelength, eq 2.1) must correspond to the difference between the ground state and possible excited electronic states of the molecule (Leifer, 1988). For a bond to be broken, photon energy (E) must be high enough to overcome molecular bond energy. Thus, direct photolysis requires the absorption spectrum of the molecule to overlap with the spectrum of incoming radiation (i.e., the photon). Photodegradation depends on the total energy absorbed in specific wavelength regions, which, in turn, depends on incident light of the system, light-absorbing properties of the molecule, and efficiency of the molecule to utilize absorbed light (quantum yield).

The absorption of a photon is a necessary, but often insufficient condition for a molecule to undergo transformation by direct photolysis (Calvert and Pitts, 1966; Harris, 1982; Leifer, 1988). The absorbed energy must be adequate to cause transformation via bond cleavage, rearrangement, oxidation, or reduction. Then, phototransformation must compete with other possible deactivation processes to form new molecular structures. Consequently, the fraction of photon-excited molecules that undergo phototransformation (i.e., the quantum yield) is generally much less than 1 (typically <0.1 and sometimes <0.01) (Harris, 1982; Mill, 1999).

Ultraviolet absorbance is quantified by the decrease in the amount of incident light as it passes through a water sample over a specified distance or path length (U.S. EPA, 2006). When light is absorbed, the intensity of the emerging light, I_{out}, is smaller than the intensity of light entering the substance, I_{in}. The product of the molar absorption coefficient (ε), concentration (c), and path length (l) of the absorbing substance is exponential to the relative absorption (I_{out}/I_{in}) and linear to the absorbance (A) (known as the Beer–Lambert law), as follows:

$$\frac{I_{out}}{I_{in}} = 10^{-\varepsilon c l} \rightarrow \log_{10}\left(\frac{I_{in}}{I_{out}}\right) = \varepsilon c l = A \qquad (2.2)$$

When irradiation interacts with opaque materials, some radiation may be absorbed, reflected, or refracted. If the surface is semi-transparent, irradiation may also be transmitted. Transmission is the process of radiation passing through matter and UV transmittance (UVT) is the percentage of light passing through a water sample at a specified UV wavelength and path length. It is related to UV absorbance (A) at a specified wavelength and path length, as follows:

$$\%UVT = 100 \times 10^{-A} \qquad (2.3)$$

With respect to biomolecules, absorption by inorganic compounds is biologically insignificant, while unsaturated organic compounds are biological absorbers in the UV region above 220 nm. The most important UV absorbers are conjugated bonds (alternating single and double bonds), and include molecules such as pyrimidine and purine. In proteins, components that can absorb UV light include the aromatic amino acids and peptide bonds, which have a double-bond character. Peptide bonds weakly absorb below 240 nm, however, there is a peptide bond in every amino acid residue, so the overall contribution is significant for proteins (Jagger, 1967).

When UV light is absorbed in the water matrix, it is no longer available to disinfect microorganisms; thus, UV light may not be the most efficient process to disinfect water with high UV absorbance (low UVT) (U.S. EPA, 2006). For example, iron compounds in water absorb UV light, reducing UV transmittance in water and increasing the UV dose required. On the other hand, light, which is reflected, refracted, or scattered by particles, may still be available for disinfection of microorganisms in water (U.S. EPA, 2006).

1.2.3 Laws of Photochemistry

The fundamental principles relating to light absorption provide the basis for understanding photochemical transformations. Photochemistry is driven by absorption of photons; three laws describe the factors that limit photochemical processes. The first law is the Grotthuss–Draper law, which states that only light that is absorbed by a chemical can bring about photochemical change. The second law is the Stark–Einstein law, which states that the primary act of light absorption by a molecule is a 1-quantum process. That is, for each photon absorbed, only one molecule is excited. The third law states the energy of an absorbed photon must be equal to or greater than the weakest bond in the molecule. These laws suggest that molecules that do not absorb light in the available wavelengths cannot undergo photochemical reactions when exposed to those wavelengths, and that UV fluence (UV dose) is directly proportional to the total number of UV photons absorbed by a given microorganism (Bolton, 2011).

1.2.4 Ultraviolet Scattering

Light scattering is the change in direction of light propagation as a result of interaction with a particle (U.S. EPA, 2006). Each particle in a solution scatters light in unique directions and patterns. As a result, the mutual influence of scattering from particles can lead to enhancement or cancellation of the integrated scattering effect. Scattering is dependent on particle size relative to the light wavelength, shape of the particle, chemical composition, number, concentration (Bohren and Huffman, 1983), and relative refractive index between the particle and the suspending medium. Large particles (>5 μm) are dominated by forward scattering, whereas, with small particles, the scattering is equally distributed in all directions (Huber and Frost, 1998).

Particles smaller than one-tenth the wavelength of incident light, such as small viruses, short polynucleotides, and globular proteins, scatter light according to Rayleigh scattering, characterized by symmetrical scattering. For these particles or large molecules, scattering is greater at the shorter wavelengths in the UV range compared to the visible range. Bacteria scatter light by different mechanisms. They are relatively transparent and contain small particles that show Rayleigh scattering; however, they also contain larger particles that will show asymmetrical scattering and also refraction. Thus, it is difficult to predict bacterial light scattering (Jagger, 1967). Particle scattering caused by colloidal particles such as clay and finely divided organic and inorganic matter may affect UV disinfection of surface water, while scattering of wastewater effluent at wavelengths above 240 nm is not substantial (Mamane et al., 2006). Ultraviolet transmittance (Section 1.2.2) is more commonly used than UV absorbance for monitoring in wastewater treatment with the presence of total suspended solids.

1.3 Germicidal Action of Ultraviolet Light

Germicidal UV action refers to the wavelengths of UV radiation that induce photochemical reactions within cells, which are effective in bacterial inactivation. In a series of papers, Gates (1930) determined the bactericidal action spectrum with similarly shaped spectra for *Staphylococcus aureus* and *Bacillus coli*. A rapid rise was observed below 300 nm to a maximum between 260 to 270 nm, a drop to a minimum near 240 nm, and rise again below 240 nm. Absorption spectra were obtained measuring the optical path through a film of bacteria. Because of scattering of light, Gates criticized previous studies using bacteria in a suspension in a fluid medium. Gates (1930) hinted that a specific substance of the cell, possibly referring to genetic material, is responsible for cell death. Similarly, Hollaender and Claus (1936) suggested that the inactivation of bacterial cells by UV irradiation may be attributed to the nuclear material or material that controls cell division as well as the bacterial cell wall, enzymes in the organism, and the colloidal structure of the cytoplasm. Setlow and Carrier (1966) showed the formation of cyclobutane-type pyrimidine dimers in polynucleotides after exposure to UV radiation. Thus, germicidal action of UV light is a combination of a variety of damage to the bacterial cell. Section 1.3.1 discusses nucleic acid damage, Section 1.3.2 presents the DNA and protein spectral absorbance, and Section 1.3.3 demonstrates a means to obtain an action spectrum of various microorganisms and compares this to DNA absorbance.

1.3.1 Nucleic Acid Damage

The genetic code of organisms is made up of nucleic acids, either as DNA or RNA; DNA serves as the databank of life, while RNA directs metabolic processes (Crittenden et al., 2012). Nucleotide bases of the DNA are adenine, guanine, thymine, and cytosine, whereas RNA contains uracil instead of thymine (Harm, 1980). Nucleic acids are heterocyclic aromatic compounds. In DNA and RNA, pyrimidine (thymine, cytosine, and uracil) form hydrogen bonds with their complementary purines (adenine and guanine), making double-stranded DNA or RNA.

Nucleic acids show significant absorption of UV photons between 200 and 300 nm, with peak absorption at 265 nm. Only minor absorption occurs at wavelengths above 300 nm; below 200 nm, UV light cannot penetrate water. Ultraviolet absorption by cellular DNA results in damage to nucleic acids, especially in bacteria and viruses, by dimerizing adjacent thymine molecules. Formation of a thymine dimer inhibits further transcription of the cell's genetic code, preventing reproduction of the organism (Setlow and Setlow, 1967) and resulting in cell inactivation and even death (Hijnen et al., 2006; Jacobs et al., 2005; Jungfer et al., 2007; Parsons, 2004).

Dimers in DNA that can be formed from thymine (T) and cytosine (C) include T<>T, C<>T, and C<>C, and, in RNA, dimers can be formed from uracil and cytosine. Cytosine dimers absorb less than thymine in the germicidal range (Harm, 1980), and the quantum yield of T<>T formation is greater than for the other dimers C<>C and C<>T (Patrick and Rahn, 1976). Organisms rich in thymine (found only in DNA) tend to be more sensitive to UV irradiation. Therefore, the relative efficiency per dinucleotide frequency for forming the dimers is in the following order: TT > CT > CC (Setlow and Carrier, 1966). For example, MS2 bacteriophage is a single-stranded RNA virus; viruses that have only RNA are less sensitive to UV radiation. The base composition for MS2 is equimolar for adenine, uracil, cytosine, and guanine, with slightly smaller amounts of adenine and slightly greater amounts of guanine (Strauss and Sinsheimer, 1963). Uracil absorbance (found only in RNA) is shifted to lower wavelengths than thymine, indicating that uracil absorbs less at the lower wavelengths (Davidson, 1965; Strauss and Sinsheimer, 1963).

1.3.2 Deoxyribonucleic Acid and Protein Absorbance

The wavelength dependence of genomic DNA absorption of UV light is obtained via UV-visible (UV-vis) spectroscopy. Figure 2.2 illustrates the spectral absorbance of DNA purified from *Pseudomonas aeruginosa* PAO1 and *Escherichia coli* K12 (Lakretz et al., 2010) using standard DNA-extraction techniques (Sambrook et al., 1989). The absorption spectrum of double-stranded DNA is determined both by the absorption spectra of the individual components and interactions between adjacent bases (Sutherland and Griffin, 1981). Because the nucleotide composition of DNA varies from one organism to another, the DNA absorbance spectrum varies slightly (Chen et al., 2009). Adenine, thymine, cytosine, and uracil have relative peak absorption values between 260 and 265 nm, with guanine peaking closer to 245 and 275 nm (Davidson, 1965).

When UV photons (polychromatic) are absorbed by a microbe, most of the germicidal action of UV light is caused by nucleic acid absorption because nucleic acids absorb in the range of 240 to 280 nm, which is 10 to 20 times higher per weight compared to protein. Proteins, however, can also be involved in inactivation of microorganisms by UV (Jagger, 1967). Various proteins and enzymes were also found to absorb UVB and UVC, resulting in further bacterial damage (Harm, 1980; Oguma et al., 2002; Sinha and

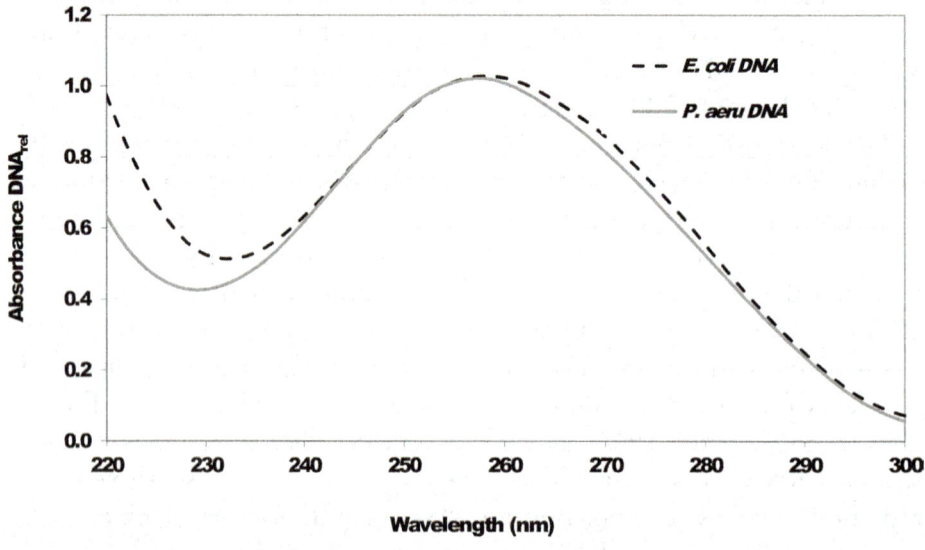

FIGURE 2.2 The DNA absorbance of *P. aeruginosa* and *E. coli* (normalized to 254 nm).

Häder, 2002). The absorption spectra of nucleic acids are very similar (Figure 2.2); however, the absorption spectra of proteins vary because of variation in the amino acid contents. Proteins typically show an absorption peak around 280 nm, with a minimum around 240 to 250 nm, while DNA shows a minimum around 230 nm. Additionally, proteins show high absorption below 230 nm because of peptide bonds (Jagger, 1967). A microbial cell, however, such as *E. coli*, contains 60 to 90% water that is transparent up to 300 nm. By dry weight, protein accounts for 50%, nucleic acids make up to 10 to 20%, and carbohydrates and lipids are each less than 10%.

1.3.3 Ultraviolet Action Spectrum

Gates (1928) discovered that there are relative biological effects of various UV wavelengths, which is the basis for action spectrum studies. A UV action spectrum is determined by measuring the dose response of a microorganism to various wavelengths and plotting the relative inactivation rate constants for the different wavelengths.

1.3.3.1 Means to Obtain an Action Spectrum

Spectral germicidal sensitivity is most often obtained by evaluating inactivation of a specific organism across wavelengths in the germicidal range. The UV inactivation rate constant is generally determined by dose-response data in the first order region and is fitted using a linear regression approach. Linear inactivation curves characterized by first order inactivation throughout the entire wavelength range are characteristic of a one-hit, one-target survival curve that assumes that a single harmful event (hit) is sufficient to inactivate a biological unit (Harm, 1980). These inactivation rate constants $k(\lambda)$, or UV sensitivity coefficients, are a measure of the sensitivity of a microorganism at a particular wavelength and, when plotted as a function of wavelength, represent the action spectrum of a microorganism (Beck et al., 2014; Cabaj et al., 2001, 2002; Mamane-Gravetz et al., 2005; Munakata et al., 1986; Rauth, 1965). The action spectrum is typically reported relative to 254 nm, the principal output of the low-pressure mercury vapor lamp. The $k(\lambda)$ from linear regression is transformed to krel(l), the relative coefficient, by dividing $k(\lambda)$ by $k(254)$.

Specific wavelengths can be obtained by using band-pass filters placed in the polychromatic light path to transmit a well-defined band of light from a polychromatic medium-pressure light source, UV light-emitting diodes (LEDs), or monochromatic sources such as lasers. Some studies used sophisticated sources as tunable laser systems, LEDs, and xenon lamps with a monochromator (Beck et al., 2014; Besaratinia et al., 2011; Freeman et al., 1989; Mamane-Gravetz et al., 2005; Wang et al., 2005). Lasers are particularly important in photochemical research because their stimulated emission produces light that is monochromatic, coherent, and intense (Wardle, 2009). The use of a laser system to generate tunable radiation of varying monochromatic wavelengths in the UV spectrum enables one to more accurately investigate the action spectrum of microorganisms.

1.3.3.2 Action Spectrum by Organism

The germicidal action spectrum of UV light is highly dependent on the wavelength and may vary by microorganism. The relative effectiveness as a function of wavelength, however, has typically been determined for *E. coli* (Meulemans, 1987). Action spectra of single-stranded DNA (ssDNA), double-stranded DNA (dsDNA), and double-stranded RNA (dsRNA) viruses as phiX174 (ssDNA), MS2 (ssRNA), T2 (dsDNA), Reovirus 3 (dsRNA), T7 coliphage (dsDNA), and T1UV coliphage (dsDNA) show a local minimum between ~232 and 240 nm and a local maximum between ~260 and 265 nm (Malley et al., 2004; Mamane-Gravetz et al., 2005; Rauth, 1965; Rontó et al., 1992) (Figure 2.3).

Other organisms such as *E. coli*, *Salmonella typhimurium*, *P. aeruginosa*, *B. subtilis*, *Cryptosporidium parvum* oocysts, and *Bacillus pumilus* also have relative peak sensitivities in the 260- to 265-nm region (also reported to extend to 270 nm, however) (Chen et al., 2009; Gates, 1930; Lakretz et al., 2010; Linden et al., 2001; Mamane-Gravetz et al., 2005; Verhoeven et al., 2013; Wang et al., 2005). Herpes simplex virus exhibited relative peak sensitivities in the region of 270 to 280 nm and thus did not follow the absorption spectrum of either nucleic acid or protein (Detsch et al., 1980). An unusual action spectrum was obtained for adenovirus 2; this spectrum was relatively flat from 240 to 260 nm, but peaked at 270 nm and rose rapidly below 240 nm for increased action at lower wavelengths. The effect of low wavelengths on adenovirus disinfection is not accounted for using standard DNA absorption weighting (Beck et al., 2014). The action spectrum of *B. pumilus* spores shows promise as a surrogate to closely match the increased action of Adenovirus at low wavelengths (Verhoeven et al., 2013), although more work needs to be done on *B. pumilus* spores.

1.3.3.3 Can the Deoxyribonucleic Acid Absorbance Spectrum Represent the Action Spectrum?

Photochemical reactions for inactivation are mostly efficient with wavelengths close to the maximum absorbance of pyrimidine (thymine, cytosine, and uracil) and purine (adenine and guanine) nucleobases (Ravanat et al., 2001). However, the absorbance spectrum of DNA isolated from various

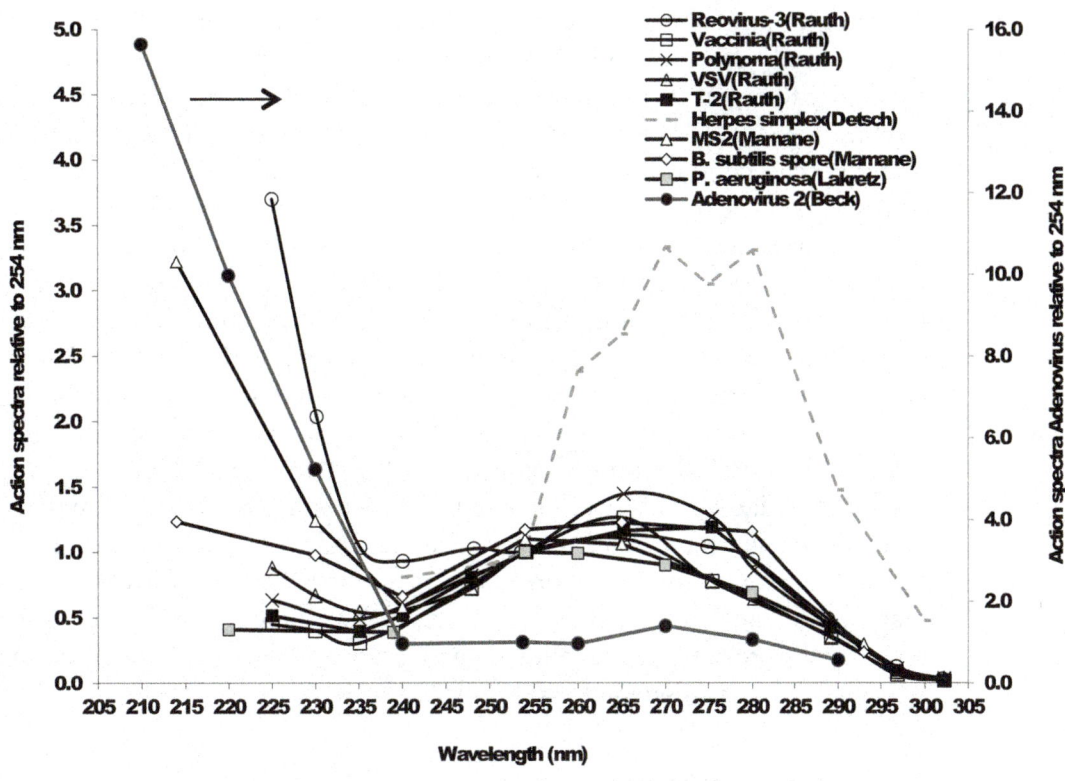

FIGURE 2.3 The UV action spectra for viral plaque-forming ability for various viruses relative to 254 nm. Data are adapted and modified from Beck et al. (2014), Detsch et al. (1980), Lakretz et al. (2010), Mamane-Gravetz et al. (2005), and Rauth (1965).

organisms can differ from the actual action spectrum because photons can be absorbed in other cell components, as shown in research with *B. subtilis* spores (Chen et al., 2009; Mamane-Gravetz et al., 2005). Thus, calculating a medium-pressure UV dose based on DNA absorption does not account for non-DNA-based damages. Beck et al. (2014) observed loss of viral infectivity greater than the loss of DNA amplification at wavelengths below 240 nm, which implied damage to viral components other than the viral genome, such as viral proteins, which play an integral role in the infection process. Moreover, isolating DNA before exposure to UV light does not enable investigation of interactions between DNA and proteins that could occur in vivo. Thus, the absorption spectrum cannot, as a rule, be substituted for the appropriate action spectrum to represent the germicidal effectiveness of given wavelengths of UV light.

1.4 Microbial Repair and Regrowth

Many microorganisms have evolved with the capability to repair UV-induced nucleic acid damage as a survival mechanism. Because UV disinfection targets nucleic acids, resulting in genomic damage, and does not primarily affect the physical integrity of a microorganism, if it has capability to repair that damage, it can restore its infectivity. Therefore, the potential exists for UV-disinfected microorganisms to repair and regrow under certain conditions. Knowledge of this potential and the conditions under which it can occur can direct engineering design to minimize any effect on overall disinfection efficacy. Repair typically is minimized or negligible when an adequate UV dose is delivered. In Germany, UV systems for disinfection of drinking water must exhibit disinfection potential corresponding to a UV dose of at least 40 mJ/cm^2 based on the work of Hoyer (1998), who illustrated that this dose is enough to eliminate the possibility of a repair process occurring that would affect the inactivation of target pathogens. The "10 States Standards" (GLUMRB, 2007) note that, for activated sludge secondary effluent, a UV dose not less than 30 mJ/cm^2 can be used, accounting for quartz sleeve fouling, and end-of-lamp-life output. Pending upstream treatment processes (e.g., filtration), even lower doses may be adequate to meet regulatory targets. Two types of microorganism repair will be reviewed here: photo-reactivation and dark repair. Regrowth issues will also be discussed.

1.4.1 Photo-Reactivation

Photo-reactivation refers to repair mechanisms initiated by the protein photolyase, which is activated by exposure to wavelengths that are 350 to 450 nm (Sancar, 1994). The ability of microorganisms to repair their DNA after UV disinfection has been reported (Harris et al., 1987; Kashimada et al., 1996; Lindenauer and Darby, 1994; Scheible et al., 1986; Sommer et al., 2000; Zimmer and Slawson, 2002). Photorepair plays an important role, especially in wastewater disinfection, where the water is potentially exposed to sunlight upon discharge to a waterbody after UV treatment. These factors imply that, to correctly evaluate the efficacy of a UV disinfection system, the possibility of subsequent repair should be taken into account.

Photorepair has been shown to occur in many waterborne bacteria, yet varies greatly between and within different species and strains (Harm, 1980). Whitby et al. (1984) reported a 0.6- to 1-log increase in coliform bacteria following photo-reactivation of UV-disinfected wastewater effluent; however, in the receiving stream, no photo-reactivation was detected (Whitby et al., 1985). Oguma et al. (2002) reported 2.1-log photorepair for *E. coli* K12 after a UV dose of 6 mJ/cm^2, while Kashimada et al. (1996) did not detect any photorepair for the same strain after a dose of 21 mJ/cm^2, indicating the importance of dose in affecting repair potential. In the case of a different *E. coli* strain, Zimmer and Slawson (2002) reported 2.6-log repair, while Sommer et al. (2000) and Harris et al. (1987) observed photorepair as high as 3.6 and 3.3 log after 8 mJ/cm^2 of UV irradiation at 254 nm. While viruses on their own cannot undergo photorepair, some types of phage have been shown to undergo repair in the presence of their host (Rodriguez et al., 2014). *Giardia* and *Cryptosporidium* both have the genomic potential for photorepair, but exhibited no repair following UV disinfection at low doses of less than 5 mJ/cm^2 (Oguma et al., 2001; Shin et al., 2001; Shin et al., 2009) or at practical doses of up to and above 16 mJ/cm^2 (Linden et al., 2002). All photorepair studies indicate that an increase in UV dose beyond the required dose for inactivation will minimize the potential for photo-reactivation events. While there is no regulated metric for including additional dose to account for reactivation in a design, application of a UV dose that achieves 1 extra log of inactivation should be sufficient to avoid reactivation, whether low-pressure high-output (LPHO) or medium-pressure UV is being installed. Because each water and system are different, this point could be tested during piloting by leaving a UV-exposed sample out in the sunlight for 4 hours before assaying the colony-forming units.

1.4.2 Dark Repair

Deoxyribonucleic acid repair mechanisms other than photorepair, such as excision repair, are named dark repair because they can repair the damaged DNA without light. In addition to excision repair, these repair processes can be initiated by recombinatorial repair and inducible error-prone repair. A review of nucleic acid repair processes is covered by Harm (1980). Dark repair processes are less effective and much slower than photorepair (Sinha and Hader, 2002) and are harder to quantify during typical inactivation studies because any dark repair process that may occur during the time of the microbial assay (on the order of days to weeks, depending on the organism) is inherently accounted for in the viability assay results. The effect of dark repair is, therefore, not necessary to account for in the design of disinfection systems.

1.4.3 Regrowth

Because UV disinfection does not leave a residual dose following the disinfection system, any surviving organisms or those only partially damaged could potentially regrow. The presence of proper nutrients would be a prerequisite for regrowth, and the potential for this to occur is site-specific. Regrowth of indicator or pathogenic bacteria can occur under the right conditions, but regrowth of viruses or protozoa would not be possible without the presence of a specific host organism and, therefore, is unlikely.

2.0 ULTRAVIOLET DOSE RESPONSES OF PATHOGENS AND SURROGATES

2.1 Collimated Beam Testing

Accurately defining the UV inactivation of microorganisms is a cornerstone of effective UV disinfection design. The most reliable method to establish dose-response relationships for microbe inactivation is use of an apparatus called a *collimated beam*.

2.1.1 Types of Collimated Beams

A collimated beam apparatus is a bench-scale UV disinfection system designed to protect the user from harmful UV rays while providing an optimal setup for proper measurement of UV-light irradiance. There is no single way to design a collimated beam apparatus, but typical systems consist of a lamp horizontally suspended above an irradiation area such that a dish of water can be irradiated. Figure 2.4 presents a few ideas for designs of collimated beam systems. Bolton and Linden (2003) detail the design and use of collimated beam systems and the specific procedures for proper calculation of the average UV irradiance in water exposed to UV light in a collimated beam apparatus. Collimated beam systems can house any type of UV lamp, but the most common type used in these systems is a low-pressure mercury vapor lamp.

2.1.2 Ultraviolet Dose Measurements and Calculation

When performing bench-scale collimated beam experiments, it is essential to accurately measure UV irradiance on the surface of the exposure dish using calibrated instrumentation that properly calculates the dose delivered

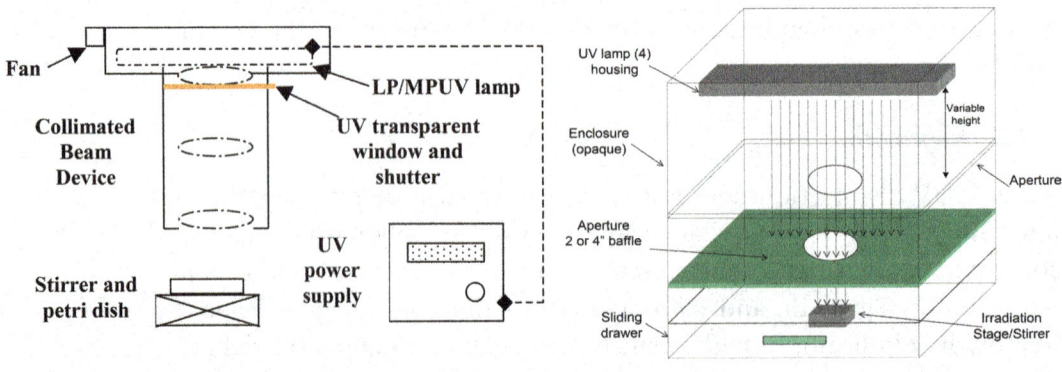

FIGURE 2.4 Types of collimated beam systems.

to the microorganisms. Ultraviolet irradiance can be measured using physical, chemical, or biological methods.

Physical methods most often use radiometers equipped with specialized detectors calibrated to read accurately at specific wavelengths. Radiometer detectors are best used to measure UV irradiance impinging normally to the surface of a water sample. Therefore, they are typically used in laboratory work where UV light can be effectively collimated. The reading from a radiometer needs to be corrected for water absorption and the depth of the liquid sample to provide a measure of the average irradiance in a batch sample. This approach is outlined in this section and is also detailed in work by Bolton and Linden (2003). In full-scale systems, a radiometer and detector system can be used to monitor the UV lamp output over time and indicates when the intensity decreases because of fouling or aging of the lamps.

Chemical actinometry is a method of measuring the UV dose of light exposure by the change produced in a photosensitive chemical. Some common actinometers used to measure UV irradiance include potassium ferrioxalate, potassium iodide/iodate, and uridine. Jin et al. (2006) reviewed the benefits and challenges of chemical actinometers with monochromatic and polychromatic UV sources. Ideal actinometers for measuring polychromatic germicidal UV light would have the following characteristics (von Sonntag and Schuchmann, 1992):

- Absorb only germicidal UV light,
- Possess an absorbance spectrum similar to that of DNA,
- Maintain constant quantum yield throughout the germicidal wavelength range,
- Be easy to quantify using standard laboratory equipment, and
- Be nontoxic to the environment.

Chemical actinometers can also be useful for checking the calibration of physical instruments. This technique is also typically used in bench-scale approaches, although full-scale dosimetry evaluations have been conducted with actinometry (Harris et al., 1987; Schulz et al., 2003).

Biological measurement of fluence has centered on the bioassay, or biodosimetry technique, for assessing the UV dose in a flow-through UV disinfection system. This technique is discussed in detail in Chapter 3. In the bioassay approach, a UV dose-response curve is established for a microbe in a collimated beam apparatus, typically using radiometry for dose measurements. A concentrated spike of the same microbe is then injected to a UV disinfection system and the log inactivation is determined based on upstream and downstream microbe concentrations. This log-inactivation value is then used, with the initial UV dose-response curve, to estimate the UV dose in the disinfection system to achieve the resulting inactivation response. This method provides a single average UV dose value (called the *reduction equivalent dose*) to the UV disinfection system and is the basis of the reactor validation protocols detailed in guidance documents such as *Technical Standard W 294: UV Systems for Disinfection in Drinking Water Supplies—Requirements and Testing* (DVGW, 1997), the *Ultraviolet Disinfection Guidelines for Drinking Water and Water Reuse* (NWRI and WRF, 2012), and *Ultraviolet Disinfection Guidance Manual for the Final Long Term 2 Enhanced Surface Water Treatment Rule* (U.S. EPA, 2006). *Bacillus subtilis* spores (Qualls and Johnson, 1983) and MS2 coliphage (Braunstein et al., 1996) are typical microbes used in bioassays in the field.

The UV dose is calculated as

$$D = E_{avg} \times t \tag{2.4}$$

Where
D = UV dose delivered to the batch system (mJ/cm^2),
E_{avg} = the average irradiance in a completely mixed batch system (mW/cm^2), and
t = time (s).

2.1.3 Factors Affecting Ultraviolet Dose Calculation

Once a measure of the incident UV irradiance is made in a collimated beam, the UV dose can be calculated using the following formula (Bolton and Linden, 2003):

$$E_{avg} = E_o \times \text{Petri Factor} \times \text{Reflection Factor} \times \text{Water Factor} \times \text{Divergence Factor} \tag{2.5}$$

where E_o is the irradiance incident on the sample surface (mW/cm^2) (the other factors are defined in the rest of this section). The irradiance incident on the water is measured by a radiometer detector, typically at the center of the irradiation dish. A number of corrections are required to obtain the average irradiance and thus the UV dose received by the water.

The Petri factor is a correction made to account for any nonhomogeneity of the irradiance field across the surface of the exposure dishwater. The *Petri factor* is defined as the ratio of the average of the incident irradiance over the area of the Petri dish to the irradiance at the center of the dish. Therefore, the Petri factor is used to correct the irradiance reading at the center of the Petri dish to represent the average incident irradiance over the surface area. A well-designed collimated beam apparatus should result in a Petri factor greater than 0.9 (90%).

The reflection factor is used because whenever a beam of light passes from one medium to another and the refractive index changes, a portion of the beam is reflected off the interface between the media. For air and water, $R = 0.025$, and the reflection factor is, therefore, $(1 - R) = 0.975$, representing the fraction of the incident light that enters the water for disinfection.

The water factor is derived from an integration of the Beer–Lambert Law over the sample depth in a completely mixed sample (Morowitz, 1950). If the water absorbs UV light, then the irradiance will decrease because of absorption as the light passes through the water. The water factor is defined as follows:

$$\text{Water Factor} = \frac{1 - 10^{-al}}{al \ln(10)} \qquad (2.6)$$

Where
- a = the decadic absorption coefficient (cm^{-1}) or absorbance for a 1-cm path length and
- l = the vertical path length (cm) of the water in the Petri dish.

The divergence factor accounts for the fact that light rays in a collimated beam are not perfectly parallel and will diverge as they pass through the water column, potentially exiting before they complete the depth of water. When the water is more than about 4 times the distance from the lamp as the aperture diameter where the beam exits the lamp housing, irradiance falls off as the inverse square of the distance from the UV lamp to the surface of the water (refer to Bolton and Linden [2003] for a mathematical explanation of this factor).

2.2 Data on Microbe Dose Response

Various dose requirements have been published for specific microorganisms and classes of microorganisms. In the United States, there are specific laws that have been promulgated for dose requirements for drinking water treatment; for wastewater, however, there are no specific doses required by national regulations. Ultraviolet dose requirements for 1- to 4-log disinfection of some pathogens and regulatory important indicator organisms are summarized in Table 2.1. Ultraviolet disinfection can often meet fecal and total coliform wastewater effluent discharge requirements at doses under 25 mJ/cm^2; various states have required doses and the 10 States Standards (GLUMRB, 2007) recommends a minimum dose of 30 mJ/cm^2. These discharge doses are substantially lower than what is required for reuse applications that meet requirements such as California's Water Recycling Criteria,

TABLE 2.1 Ultraviolet dose requirements for 1- to 4-log credit disinfection of some pathogens and regulatory-important indicator organisms (UV dose [fluence] provided in mJ/cm^2).

Pathogen	1 log	2 log	3 log	4 log
Giardia[a]	2.1	5.2	11	22
Cryptosporidium[a]	2.5	5.8	12	22
Viruses[a]	58	100	143	186
E. coli[b]	3.4	4.7	5.8	7.0
Fecal Coliform[c]	4.6	9.2	13.8	18.4
Total Coliform[c]	2.0	4.0	5.9	7.9

[a]From U.S. EPA (2006)
[b]From Chevrefils et al. (2006) (taken as an average of all the *E. coli* data presented)
[c]From Hijnen et al. (2006) (from Figure 8)

Title 22, for which a UV dose of 100 mJ/cm^2 is required for media filtered water (it is important to note that, following membrane filtration, doses are lower [NWRI and WRF, 2012]). For a review of the literature regarding the dose-response of numerous pathogens and surrogates of interest in disinfection processes, refer to Hijnen et al. (2006). Details for specific microorganism classes are discussed in the following subsections.

2.2.1 Bacteria

Escherichia coli typically is the test bacterium in UV disinfection studies because of its widespread use as a fecal indicator (U.S. EPA, 2006). Chevrefils et al. (2006) published an impressive summary on the UV dose response of various organisms including bacteria, categorized to bacterium type, lamp type, and UV dose per 1- to 6-log reduction. Required doses for 4-log inactivation using low-pressure lamps ranged from 5 to 10 mJ/cm^2 for *E. coli*; there was similar sensitivity for most of the other bacteria examined as *Aeromonas, Legionella pneumophila, Salmonella typhimurium,* and *Streptococcus faecalis*.

2.2.2 Viruses

Enteric viruses (enteroviruses), such as adenovirus, norovirus, rotavirus, and hepatitis A, frequently occur in water and wastewater (Abbaszadegan et al., 2003; Hamza et al., 2009; Wong et al., 2009). Enteroviruses reproduce in the gastrointestinal tract after infection occurs. Bacteriophages (phage) are viruses that infect host bacteria and are often used as human virus surrogates. Viruses are more resistant to disinfection than fecal bacteria, with different inactivation kinetics; thus, viral inactivation should be considered in wastewater treatment and reclamation (Tree et al., 1997).

Most of the published data on fundamental virus inactivation kinetics were conducted at bench scale with seeded viruses. Chevrefils et al. (2006) summarized more than 50 publications on viruses, categorized to virus type, host, lamp type, and UV dose. Required doses for 4-log inactivation using low-pressure lamps ranged from 90 to 120 mJ/cm^2 for MS2, 7 to 10.5 mJ/cm^2 for PhiX 174, 22 to 47 mJ/cm^2 for Poliovirus 1, 16 to 30 mJ/cm^2 for Hepatitis A, and 112 to 165 mJ/cm^2 for Adenovirus (types 2, 15, 40, and 41). The enteroviruses as a group show little variability in resistance to UV light, except for adenovirus type 2 (Gerba et al., 2002). Adenoviruses were more resistant than other enteric viruses because they contain dsDNA and are able to use the host cell enzymes to repair damages in the DNA caused by UV irradiation.

2.2.3 Protozoa

Before 1998, it was thought that UV irradiation was not effective for inactivation of protozoan pathogens (Campbell et al., 1995). However, the studies upon which those conclusions were reached used a form of viability analysis for protozoa evaluation. Clancy et al. (1998, 2000) used animal infectivity assay as a means to evaluate UV disinfection performance and discovered that UV treatment was effective at inactivating *Cryptosporidium*. Other studies followed that work, verifying that the required UV dose for inactivation of *Cryptosporidium parvum* was very low, reportedly below 10 mJ/cm^2 for greater than 4-log inactivation (Clancy et al., 2004; Shin et al., 2001). Inactivation of *Cryptosporidium hominis* was determined to be similar to that of *C. parvum* (Johnson et al., 2005). Similar studies with Giardia cysts using animal infectivity testing illustrated that low doses of UV light (less than 5 mJ/cm^2) resulted in high levels of inactivation (Linden et al., 2002; Shin et al., 2009). Another protozoan pathogen that has been studied is *Toxoplasma gondii*, which was also found to be similarly susceptible to inactivation by UV light (Dumetre et al., 2008; Ware et al., 2010). Thus, while wastewater regulations do not specify inactivation of protozoa, minimizing the transmission of these pathogens through the urban water cycle requires effective disinfection at the water resource recovery facility, and UV light is an effective mechanism of inactivating protozoan pathogens at low doses.

3.0 ULTRAVIOLET LAMP TECHNOLOGIES

There are two main types of UV lamps commonly used in engineered disinfection applications; these are both based on mercury vapor discharge.

Mercury inside the lamp is excited by applying a voltage to the gas mixture, by which the mercury atoms are excited to the first energy state above the ground state. The transition of electrons back to a ground state results in emission of electromagnetic energy (photons) at wavelengths characteristic of the mercury. The two lamp types are commonly referred to as *low-pressure* and *medium-pressure mercury vapor UV lamps*; emission spectra of low-pressure and medium-pressure UV lamps are shown in Figure 2.5.

3.1 Low-Pressure Ultraviolet Lamps

The low-pressure UV lamp is a mercury-vapor lamp that operates at a relatively low internal mercury vapor pressure of 0.13 to 1.3 Pa (2×10^{-5} to 2×10^{-4} psi) and a low lamp temperature (around 40 to 80 °C). This operation results in approximately 85% of the UV light emitted being monochromatic at 253.7 nm, with other small peaks above 300 nm and another significant

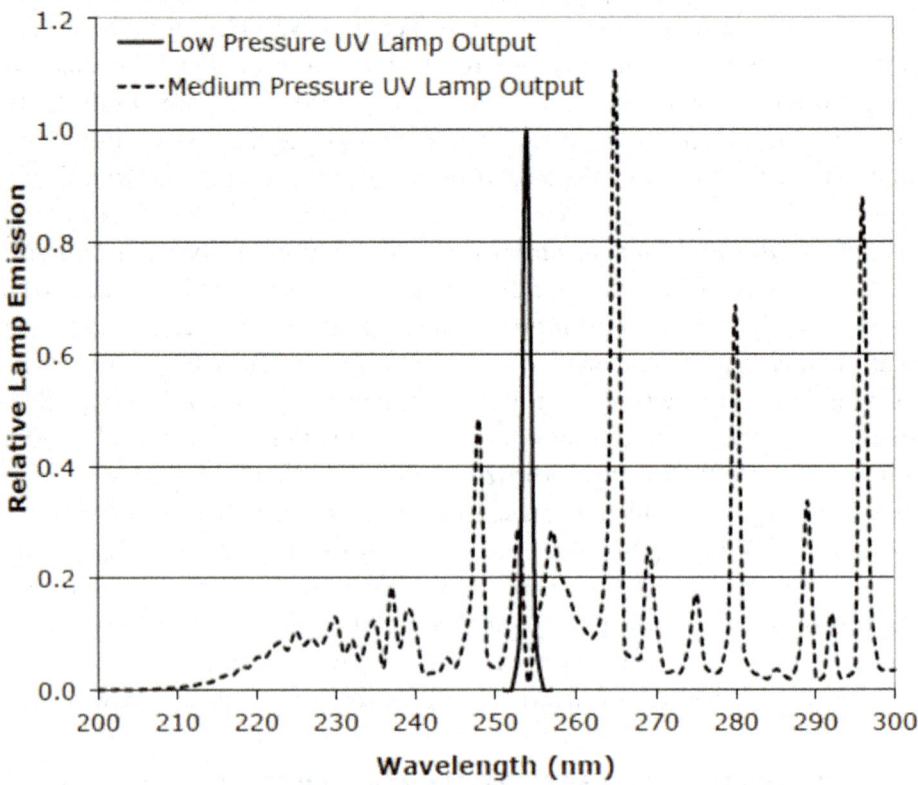

FIGURE 2.5 Emission spectra of low- and medium-pressure mercury vapor lamps. Note that the low-pressure output is relevant for low-pressure and LPHO lamp systems.

line emission at 185 nm. These light sources are tubular in construction, and are similar in design and operation to household fluorescent lamps. Conventional low-pressure lamps (sometimes referred to as *low-pressure low-output* [LPLO]) are typically used for smaller wastewater flows, where the number of lamps can be minimized and where UV transmittance is not a limiting factor. Output from LPLO lamps can be as low as a few watts and is typically up to about 150 W per lamp.

3.2   Low-Pressure High-Output Ultraviolet Lamps

Low-pressure high-output lamps are similar to the low-pressure mercury vapor lamp, but operate under higher electrical input, resulting in a higher UV intensity output. Similarly, the lamps also have essentially monochromatic light output at 254 nm. To control the vapor pressure under the higher energy input, the LPHO lamps contain solid spots of a mercury amalgam positioned on the inside of the lamp quartz tube. These amalgams are a mercury alloy with an element such as gallium or indium. This amalgam helps to control the mercury vapor pressure, allowing a characteristic low-pressure output at 253.7 nm to be emitted at a higher intensity than is possible without the amalgam. Recent innovations in LPHO technology have led to the development and commercialization of LPHO lamps with input up to 1000 W per lamp.

3.3   Medium-Pressure Ultraviolet Lamps

The medium-pressure lamp is also a mercury vapor lamp, but it operates at an internal pressure of 1.3 to 13 000 Pa (2 to 200 psi) and a higher temperature (600 to 900 °C) and electrical input than low-pressure lamps. This results in a polychromatic (or broad-spectrum) output of UV light, including output at wavelengths outside of the germicidal range. The germicidal efficiency of medium-pressure lamps is approximately 10 to 15%. Thus, medium-pressure lamps are less germicidally efficient than low-pressure lamps; however, because the UV output is much greater, they emit approximately 10 to 50 times the germicidal UV output of low-pressure lamps and 4 to 10 times the germicidal output of LPHO lamps. The higher UV output results in a more compact design than most LPHO systems, with fewer lamps for the equivalent effect.

3.4   Alternative Lamp Technologies

One challenge of conventional lamp technology is disposal of lamps that have mercury. Some lamp manufacturers recycle these materials, which has driven research into other methods of generating UV light for disinfection applications. Additional research has also focused on development of UV sources that are more energy efficient or have longer lamp life.

3.4.1 Ultraviolet Light-Emitting Diodes

Ultraviolet LEDs are small (5 to 9 mm in diameter) and do not contain glass, filament, or mercury, which helps their transport and disposal (Bettles et al., 2007). Warm-up time is not required for LEDs, saving energy and allowing intermittent use and quick recovery from a power failure. Light-emitting diodes are replacing a number of light sources currently used today, including traffic lights and household lights. Light-emitting diodes have an excellent track record for lowering system costs through energy savings, lower maintenance, and longer replacement intervals. The average electrical-to-germicidal efficiency of low-pressure UV mercury tube lamps is 35 to 38% (U.S. EPA, 2006). Visible LEDs can operate at 75% efficiency for 10 years (100 000 hours) (Bettles et al., 2007). Currently, the efficiencies of UV LEDs are less than 1%, with lifetimes of around 1000 hours (Bettles et al., 2007; Gaska, 2007), which leads to increased operation and maintenance and life cycle replacement costs. The availability of specific output wavelengths using UV LEDs may increase their inactivation efficacy. Ultraviolet LEDs currently operate in the wavelength range of 247 to 365 nm (Gaska, 2007). The inherent low power of UV LEDs (on the order of milliwatts per centimeter squared) make application of these light sources most relevant for small systems.

3.4.2 Microwave Ultraviolet Radiation

Microwave radiation is non-ionizing electromagnetic radiation with wavelengths ranging from 1.0 mm to 1.0 m between infrared and radio waves (frequencies from 300 to 0.3 GHz, respectively). The lower band of the ultra high-frequency band is between 300 and 900 MHz. The energy of microwave photons ($E = 0.98$ J/mol at $v = 2.45$ GHz; typical for domestic and industrial microwave ovens) is insufficient to disrupt bonds of organic molecules and considerably lower than the energy of UV-vis radiation ($E = 600$ to 170 kJ/mol at $v = 200$ to 700 nm), which is responsible, directly or indirectly, for photodegradation of organic molecules (Církva and Relich, 2011). Microwave effects can be thermal or nonthermal, both of which result in contaminant degradation (Remya and Lin, 2011) and pathogen inactivation (Tyagi and Lo, 2013).

One use of microwave radiation in water treatment is in the excitation of UV lamps. Conventional UV lamps containing mercury (Hg) vapor (as low-pressure, LPHO, and medium-pressure lamps) are excited by electrical discharges sustained between electrodes; as a result, various electrode effects such as sputtering and evaporation limit the lamp lifetime. Deposition of electrode material on the lamp envelope will darken it, reduce light output, and add to mercury consumption. The issue of lamp lifetime caused by decay of electrodes has been addressed by the use of electrodeless lamps, some of which are microwave-excited. Microwave-excited electrodeless lamps generating UV-vis light is an emerging field of practice. These lamps consist of a transparent envelope filled with an inert (noble) gas and an excitable substance (metal or gas) sealed under pressure. Based on the composition of the lamp filling, it is possible to obtain a wide range of emission spectra (Nascimento and Azevedo, 2013). As with electrode discharge lamps, low-pressure lamps tend to have narrow line radiation, while the lines can be considerably broadened in high-pressure operation. Microwave radiation can be generated from various sources, although most typically from magnetrons (Církva and Relich, 2011), and is directed through wave guides into the quartz lamp sleeves containing the gas filling.

Gutierrez et al. (2006) highlighted advantages and disadvantages of microwave electrodeless UV lamps as detailed by Phillips (1983). Advantages included quick warm up time and shut-off, increased lamp life, eliminating electrodes allows narrower lamps, radiation produced throughout the entire lamp length, adaptable shapes, and simple maintenance (Horikoshi et al., 2002). Disadvantages include additional components, limited magnetron lifetime, and overheating that can cause failure in spectral emission (Nascimento and Azevedo, 2013). Microwave lamp systems have been investigated for direct degradation of contaminants in water (Remya and Lin, 2011; Ta et al., 2006; Zhang et al., 2006) or along with a photocatalyst as TiO_2 and/or oxidants as H_2O_2 for enhanced oxidation (Horikoshi et al., 2002; Nascimento and Azevedo, 2013). For applications using argon-mercury microwave lamps, only a few studies have been published (Al-Shamma'a et al., 2001; Barkhudarov et al., 2009; Barkhudarov et al., 2012; Christofi et al., 2008). Barkhudarov et al. (2012) presented a low-pressure mercury-argon vapor mixture using a microwave source for excitation, with generation of UV and ozone by using both resonance mercury lines at $\lambda 1 = 185$ nm and $\lambda 2 = 254$ nm for additional bacterial inactivation activity (Christofi et al., 2008).

3.4.3 Pulsed Ultraviolet Lamps

PUV lamps are mercury-free and do not require a warm-up period (instant-on). A high-power electrical pulse is discharged in micro-second bursts to produce intense light pulses; the discharge is in a nontoxic rare gas (e.g., xenon or krypton). Pulsed lamps can be either of the flashlamp type or a surface discharge type. Surface discharge lamps use a plasma discharge on the outside of a dielectric material enclosed within a tube with a large diameter (Schaefer, 1999, 2004), whereas flashlamps use discharges that are constrained within a relatively small envelope. Bohrerova et al. (2008) examined the use of PUV lamps in disinfection and found that inactivation of bacteria and phage was significantly faster using PUV irradiation compared to low- or medium-pressure UV lamps at equivalent UV doses. A significant fraction of enhanced PUV inactivation efficiency was attributable to wavelengths greater than 295 nm.

3.4.4 Excimer Lamps

Excimer lamps are mercury-free UV sources based on the transition of rare gas excited dimers, halogen excited dimers, or rare gas halide excited complexes, and emit nearly monochromatic radiation at wavelengths ranging from 172 to 345 nm (Oppenländer, 2007; Sosnin et al., 2006). For example, XeBr excilamps emit at 282 nm, KrCl lamps emit at 222 nm, and KrBr lamps emit at 206 nm. These wavelengths may have advantages over conventional UV sources for specific microorganisms or under certain water quality conditions.

4.0 REFERENCES

Abbaszadegan, M.; Lechevallier, M.; Gerba, C. (2003) Occurrence of Viruses in US Groundwaters. *J. Am. Water Works Assoc.*, **95** (9), 107–120.

Al-Shamma'a, A. I.; Pandithas, I.; Lucas, J. (2001) Low-Pressure Microwave Plasma Ultraviolet Lamp for Water Purification and Ozone Applications. *J. Phys. D Appl. Phys.*, **34** (18), 2775.

Barkhudarov, E. M.; Denisova, N. V.; Kossyi, I. A.; Misakyan, M. A. (2009) Resonance Microwave Discharge as a Source of UV Radiation. *Plasma Physics Rep.,* **35** (7), 559–566.

Barkhudarov, E. M.; Kozlov, Y. N.; Kossyi, I. A.; Malykh, N. I.; Misakyan, M. A.; Taktakishvili, I. M.; Khomichenko, A. A. (2012) Electrodeless Microwave Source of UV Radiation. *Tech. Physics,* **57** (6), 885–887.

Beck, S. E.; Rodriguez, R. A.; Linden, K. G.; Hargy, T. M.; Larason, T. C.; Wright, H. B. (2014) Wavelength Dependent UV Inactivation and DNA Damage of Adenovirus as Measured by Cell Culture Infectivity and Long Range Quantitative PCR. *Environ. Sci. Technol.*, **48** (1), 591–598.

Belliveau, B. H.; Beaman, T. C.; Pankratz, H. S.; Gerhardt, P. (1992) Heat Killing of Bacterial Spores Analyzed by Differential Scanning Calorimetry. *Bacteriology,* **174**, 4463–4474.

Berney, M.; Hammes, F.; Bosshard, F.; Weilenmann, H.-U.; Egli, T. (2007) Assessment and Interpretation of Bacterial Viability by Using the LIVE/DEAD BacLight Kit in Combination with Flow Cytometry. *Appl. Environ. Microbiol.*, **73**, 3283–3290.

Berney, M.; Weilenmann, H.-U.; Simonetti, A.; Egli, T. (2006) Efficacy of Solar Disinfection of *Escherichia coli, Shigella flexneri, Salmonella typhimurium* and *Vibrio cholerae. Appl. Microbiol.*, **101**, 828–836.

Besaratinia, A.; Yoon, J. I.; Schroeder, C.; Bradforth, S. E.; Cockburn, M.; Pfeifer, G. P. (2011) Wavelength Dependence of Ultraviolet Radiation-Induced DNA Damage as Determined by Laser Irradiation Suggests that *Cyclobutane pyrimidine* Dimers Are the Principal DNA Lesions Produced by Terrestrial Sunlight. *FASEB J.*, **25** (9), 3079–3091.

Bettles, T.; Schujman, S.; Smart, J. A.; Liu, W.; Schowalter, L. (2007) UV Light Emitting Diodes—Their Applications and Benefits. *Proceedings of the International Ultraviolet Association Conference;* Los Angeles, California, Aug 27–30.

Bohren, C. F.; Huffman, D. R. (1983) *Absorption and Scattering of Light by Small Particles*; Wiley-Interscience: New York.

Bohrerova, Z.; Shemer, H.; Lantis, R.; Impellitteri, C. A.; Linden, K. G. (2008) Comparative Disinfection Efficiency of Pulsed and Continuous-Wave UV Irradiation Technologies. *Water Res.*, **42** (12), 2975–2982.

Bolton, J. R. (2011) *Ultraviolet Applications Handbook,* 3rd ed.; ICC Lifelong Learn Inc.: Edmonton, Alberta, Canada.

Bolton, J. R.; Linden, K. G. (2003) Standardization of Methods for Fluence UV Dose Determination in Bench-Scale UV Experiments. *J. Environ. Eng.*, **129**, 3, 209–215.

Braunstein, J. L.; Loge, F. J.; Tchobanoglous, G.; Darby, J. L. (1996) Ultraviolet Disinfection of Filtered Activated Sludge Effluent for Reuse Applications. *Water Environ. Res.*, **68** (2), 152–161.

Cabaj, A.; Sommer, R.; Pribil, W.; Haider, T. (2002) The Spectral Sensitivity of Microorganisms Used in Biodosimetry. *Water Sci. Technol.: Water Supp.*, **2**, 175–181.

Cabaj, A.; Sommer, R.; Pribil, W.; Haider, T. (2001) What Means "Dose" in UV Disinfection with Medium Pressure Lamps? *Ozone Sci. Eng.*, **23**, 239–244.

Calvert, J. G.; Pitts, J. N., Jr. (1966) *Photochemistry;* Wiley & Sons: New York.

Campbell, A. T.; Robertson, L. J.; Snowball, M. R.; Smith, H. V. (1995) Inactivation of Oocysts of *Cryptosporidium parvum* by Ultraviolet Irradiation. *Water Res.*, **29**, 2583–2586.

Chen, R. Z.; Craik, S. A.; Bolton, J. R. (2009) Comparison of the Action Spectra and Relative DNA Absorbance Spectra of Microorganisms: Information Important for the Determination of Germicidal Fluence (UV Dose) in an Ultraviolet Disinfection of Water. *Water Res.*, **43** (20), 5087–5096.

Chevrefils, G.; Caron, É.; Wright, H.; Sakamoto, G.; Payment, P.; Barbeau, B.; Cairns, B. (2006) UV Dose Required to Achieve Incremental Log Inactivation of Bacteria, Protozoa and Viruses. *IUVA News*, **8** (1), 38–45.

Christofi, N.; Misakyan, M. A.; Matafonova, G. G.; Barkhudarov, E. M.; Batoev, V. B.; Kossyi, I. A.; Sharp, J. (2008) UV Treatment of Microorganisms on Artificially-Contaminated Surfaces Using Excimer and Microwave UV Lamps. *Chemosphere*, **73** (5), 717–722.

Církva, V.; Relich, S. (2011) Microwave Photochemistry and Photocatalysis. Part 1: Principles and Overview. *Curr. Org. Chem.*, **15** (2), 248–264.

Clancy, J. L.; Bukhari, Z.; Hargy, T. M.; Bolton, J. R.; Dussert, B. W.; Marshall, M. M. (2000) Using UV to Inactivate Cryptosporidium. *J.—Am. Water Works Assoc.*, **92** (9), 97–104.

Clancy, J. L.; Hargy, T. M.; Marshall, M. M.; Dyksen, J. E. (1998) UV Light Inactivation of *Cryptosporidium* Oocysts. *J.—Am. Water Works Assoc.*, **90**, 92–102.

Clancy, J. L.; Marshall, M. M.; Hargy, T. M.; Korich, D. J. (2004) Susceptibility of Five Strains of *Cryptosporidium parvum* Oocysts to UV Light. *J.—Am. Water Works Assoc.*, **96**, 84–92.

Crittenden, J. C.; Trussell, R. R.; Hand, D. W.; Howe, K. J.; Tchobanoglous, G. (2012) *Water Treatment: Principles and Design*, 3rd ed.; Wiley & Sons: New York.

Davidson, J. N. (1965) *Biochemistry of the Nucleic Acids*, 5th ed.; London Methuen Co. Ltd.: London.

Detsch, R. M.; Bryant, F. D.; Coohill, T. P. (1980) The Wavelength Dependence of Herpes Simplex Virus Inactivation by Ultraviolet Radiation. *Photochem. Photobiol.*, **32** (2), 173–176.

Deutscher Verein des Gas und Wasserfaches (1997) Technical Standard W 294: UV Systems for Disinfection in Drinking Water Supplies—Requirements and Testing; Deutscher Verein des Gas und Wasserfaches: Bonn, Germany.

Dumetre, A.; Le Bras, C.; Baffet, M.; Meneceur, P.; Dubey, J. P.; Derouin, F.; Duguet, J. P.; Joyeux, M.; Moulin, L. (2008) Effects of Ozone and Ultraviolet Radiation Treatments on the Infectivity of Toxoplasma Gondii Oocysts. *Vet. Parasitol.*, **153**, 209–213.

Foyer, C. H.; Lelandais, M.; Kunert K. J. (1994) Photooxidative Stress in Plants. *Physiol. Plantarum,* **92** (4), 696–717.

Freeman, S. E.; Hacham, H.; Gange, R. W.; Maytum, D. J.; Sutherland, J. C.; Sutherland, B. M. (1989) Wavelength Dependence of Pyrimidine Dimer Formation in DNA of Human Skin Irradiated In Situ with Ultraviolet Light. *Proc. Natl. Acad. Sci.,* **86** (14), 5605–5609.

Gaska, R. (2007) Deep Ultraviolet Light Emitting Diodes for Water Monitoring and Disinfection. *Proceedings of the International Ultraviolet Association Conference;* Los Angeles, California, Aug 27–30.

Gates, F. L. (1928) A Study of the Bactericidal Action of Ultraviolet Light. *J. Contam. Hydrol.,* **13**, 231–260.

Gates, F. L. (1930) A Study of the Bactericidal Action of Ultra Violet Light III. The Absorption of Ultra Violet Light by Bacteria. *J. Gen. Physiol.,* **14** (1), 31–42.

Gerba, C. P.; Gramos, D. M.; Nwachuku, N. (2002) Comparative Inactivation of Enteroviruses and Adenovirus 2 by UV Light. *Appl. Environ. Microbiol.,* **68** (10), 5167–5169.

Great Lakes-Upper Mississippi River Board (2007) 10 States Standards—Recommended Standards for Water Works. https://www.broward.org/WaterServices/Documents/states_standards_water.pdf (accessed Jan 2015).

Gutierrez, R. L.; Bourgeous, K. N.; Salveson, A.; Meir, J.; Slater, A. (2006) Microwave UV: A New Wave of Tertiary Disinfection. *Proceedings of the 79th Annual Water Environment Federation Technical Exhibition and Conference* [CD-ROM]; Water Environment Federation: Alexandria, Virginia; Dallas, Texas, Oct 21–25; pp 2853–2864.

Hamza, I. A.; Jurzik, L.; Stang, A.; Sure, K.; Uberla, K.; Wilhelm, M. (2009) Detection of Human Viruses in Rivers of a Densely-Populated Area in Germany Using a Virus Adsorption Elution Method Optimized for PCR Analyses. *Water Res.,* **43** (10), 2657–2668.

Harm, W. (1980) *Biological Effects of Ultraviolet Radiation;* Press Syndicate of the University of Cambridge: Cambridge, U.K.

Harris, G. D.; Adams, V. D.; Sorensen, D. L.; Curtis, M. S. (1987) Ultraviolet Inactivation of Selected Bacteria and Viruses with Photoreactivation of the Bacteria. *Water Res.,* **21** (6), 687–692.

Harris, J. (1982) Rates of Direct Aqueous Photolysis. In *Handbook of Chemical Property Estimation Methods: Environmental Behavior of Organic Compounds*; Lyman, W., Reehl, W., Rosenblatt, D., Eds.; McGraw-Hill: New York.

Hijnen, W. A. M.; Beerendonk, E. F.; Medema, G. J. (2006) Inactivation Credit of UV Radiation for Viruses, Bacteria and Protozoan (oo) Cysts in Water: A Review. *Water Res.*, **40** (1), 3–22.

Hollaender, A.; Claus, W. D. (1936) The Bactericidal Effect of Ultraviolet Radiation on *Escherichia coli* in Liquid Suspensions. *J. Gen. Physiol.*, **19** (5), 753–765.

Horikoshi, S.; Hidaka, H.; Serpone, N. (2002) Environmental Remediation by an Integrated Microwave/UV-Illumination Method II: Characteristics of a Novel UV–VIS–Microwave Integrated Irradiation Device in Photodegradation Processes. *J. Photochem. Photobiol. A: Chem.*, **153** (1), 185–189.

Hoyer, O. (1998) Testing Performance and Monitoring of UV Systems for Drinking Water Disinfection. *Water Supply*, **16** (1/2), 419–442.

Huber, E.; Frost, M. (1998) Light Scattering by Small Particles. *J. Water Services Res. Technol.-Aqua*, **47** (2), 87–94.

Jacobs, J.; Carroll, T.; Sundin, G. (2005) The Role of Pigmentation, Ultraviolet Radiation Tolerance, and Leaf Colonization Strategies in the Epiphytic Survival of Phyllosphere Bacteria. *Microb. Ecol.*, **49** (1), 104–113.

Jagger, J. (1967) *Introduction to Research in Ultraviolet Photobiology*; Prentice-Hall: Eaglewood Cliffs, New Jersey.

Jin, S.; Mofidi, A. A.; Linden, K. G. (2006) Polychromatic UV Fluence Measurements Using Chemical Actinometry, Biodosimetry, and Mathematical Techniques. *J. Environ. Eng.*, **132** (8), 831–841.

Johnson, A. M.; Linden, K.; Ciociola, K. M.; De Leon, R.; Widmer, G.; Rochelle, P. A. (2005) UV Inactivation of *Cryptosporidium hominis* as Measured in Cell Culture. *Appl. Environ. Microbiol.*, **71**, 2800–2802.

Jungfer, C.; Schwartz, T.; Obst, U. (2007) UV-Induced Dark Repair Mechanisms in Bacteria Associated with Drinking Water. *Water Res.*, **41** (1), 188–196.

Kashimada, K.; Kamiko, N.; Yamamoto, K.; Ohgaki, S. (1996) Assessment of Photoreactivation Following Ultraviolet Light Disinfection. *Water Sci. Technol.*, **33** (10-11), 261–269.

Lakretz, A.; Ron, E. Z.; Mamane, H. (2010) Biofouling Control in Water by Various UVC Wavelengths and Doses. *Biofouling*, **26**, 257–267.

Leifer, A. (1988) *The Kinetics of Environmental Aquatic Photochemistry. Theory and Practice. ACS Professional Reference Book*; American Chemical Society: Washington, D.C.

Linden, K. G.; Shin, G-A.; Faubert, G.; Cairns, W.; Sobsey, M. D. (2002) UV Disinfection of *Giardia lamblia* in Water. *Environ. Sci. Technol.*, **36** (11), 2519–2522.

Linden, K.; Shin, G. A.; Sobsey, M. D. (2001) Relative Efficacy of UV Wavelengths for the Inactivation of *Cryptosporidium parvum*. *Water Sci. Technol.*, **43** (12), 171–174.

Lindenauer, K. G.; Darby, J. L. (1994) Ultraviolet Disinfection of Wastewater: Effect of Dose on Subsequent Photoreactivation. *Water Res.*, **28** (4), 805–817.

Malley, J. P.; Ballester, N. A.; Margolin, A. B.; Linden, K. G.; Mofidi, A.; Bolton, J. R.; Crozes, G.; Cushing, B.; Mackey, E.; Laine, J. M.; Janex, M. L. (2004) *Inactivation of Pathogens with Innovative UV Technologies*; American Water Works Association Research Foundation: Denver, Colorado.

Mamane, H.; Ducoste, J. J.; Linden, K. G. (2006) Effect of Particles on Ultraviolet Light Penetration in Natural and Engineered Systems. *Appl. Optics*, **45** (8), 1844–1856.

Mamane-Gravetz, H.; Linden, K. G.; Cabaj, A.; Sommer, R. (2005) Spectral Sensitivity of Bacillus subtilis Spores and MS2 Coliphage for Validation Testing of Ultraviolet Reactors for Water Disinfection. *Environ. Sci. Technol.*, **39** (20), 7845–7852.

Meulemans, C. C. E. (1987) The Basic Principles of UV–Disinfection of Water. *Ozone Sci. Eng. J.*, **9** (4), 299–313.

Mill, T. (1999) Predicting Photoreaction Rates in Surface Water. *Chemosphere*, **38**, 1379–1390.

Morowitz, H. J. (1950) Absorption Effects in Volume Irradiation Dosimetry. *Science*, **111**, 229–230.

Munakata, N.; Hieda, K.; Kobayashi, K.; Ito, A.; Ito, T. (1986) Action Spectra in Ultraviolet Wavelengths (150–250 nm) for Inactivation and Mutagenesis of *Bacillus subtilis* Spores Obtained with Synchrotron Radiation. *Photochem. Photobiol.*, **44**, 385–390.

Nascimento, U. M.; Azevedo, E. B. (2013) Microwaves and Their Coupling to Advanced Oxidation Processes: Enhanced Performance in Pollutants Degradation. *J. Environ. Sci. Health, Part A*, **48** (9), 1056–1072.

National Water Research Institute; Water Research Foundation (2012) *Ultraviolet Disinfection Guidelines for Drinking Water and Water Reuse*, 3rd ed.; National Water Research Institute: Fountain Valley, California.

Oguma, K.; Katayama, H.; Mitani, H.; Morita, S.; Hirata, T.; Ohgaki, S. (2001) Determination of Pyrimidine Dimers in *Escherichia coli* and *Cryptosporidium parvum* during UV Light Inactivation, Photoreactivation, and Dark Repair. *Appl. Environ. Microbiol.*, **67** (10), 4630–4637.

Oguma, K.; Katayama, H.; Ohgaki, S. (2002) Photoreactivation of *Escherichia coli* after Low- or Medium-Pressure UV Disinfection Determined by an Endonuclease Sensitive Site Assay. *Appl. Environ. Microbiol.*, **68** (12), 6029–6035.

Oppenländer, T. (2007) Mercury-Free Sources of VUV-UV Radiation: Application of Modern Excimer Lamps (Excilamps) for Water and Air Treatment. *J. Environ. Eng. Sci.*, **6**, 253–264.

Parsons, S. A. (2004) *Advanced Oxidation Processes for Water and Wastewater Treatment*; IWA Publishing: London, U.K.

Patrick, M. H.; Rahn, R. O. (1976) Photochemistry of DNA and Polynucleotides. *Photochem. Photobiol. Nucleic Acids*, **2**, 35–91.

Patterson, D.; Gillespie, D. (1972) Effect of Elevated Temperatures on Protein Synthesis in *Escherichia coli*. *Bacteriology*, **112**, 1177–1183.

Paul, N. D.; Gwynn-Jones, D. (2003) Ecological Roles of Solar UV Radiation: Towards an Integrated Approach. *Trends Ecol. Evolution*, **18** (1), 48–55.

Pfeifer, G. P.; You, Y. H.; Besaratinia, A. (2005) Mutations Induced by Ultraviolet Light. *Mutation Res./Fundam. Molecular Mech. Mutagenesis*, **571** (1), 19–31.

Phillips, R. (1983) *Sources and Applications of Ultraviolet Radiation*; Academic Press: London, U.K.

Qualls, R. G.; Johnson, J. D. (1983) Bioassay and Dose Measurement in UV Disinfection. *Appl. Environ. Microbiol.*, **45** (3) 872–877.

Rauth, A. M. (1965) The Physical State of Viral Nucleic Acid and the Sensitivity of Viruses to Ultraviolet Light. *Biophysical J.*, **5** (3), 257–273.

Ravanat, J. L.; Douki, T.; Cadet, J. (2001) Direct and Indirect Effects of UV Radiation on DNA and Its Component. *Photochem. Photobiol.*, **63**, 88–102.

Remya, N.; Lin, J. G. (2011) Current Status of Microwave Application in Wastewater Treatment—A Review. *Chem. Eng. J.*, **166** (3), 797–813.

Rodriguez, R. A.; Bounty, S. H.; Beck, S. E.; Chan, C.; McGuire, C.; Linden, K. G. (2014) Photoreactivation of Bacteriophages after UV Disinfection: Role of Genome Structure and Impacts of UV Source. *Water Res.*, **55**, 143–149.

Rontó, G.; Gáspár, S.; Bérces, A. (1992) Phages T7 in Biological UV Dose Measurements. *J. Photochem. Photobiol. B: Biology*, **12** (3), 285–294.

Sambrook, J.; Fritsch, E.; Maniatis, T. (1989) *Molecular Cloning: A Laboratory Manual*, 2nd ed.; Cold Spring Harbor Laboratory Press: Cold Spring Harbor, New York.

Sancar, A. (1994) Structure and Function of DNA Photolyase. *Biochemistry*, **33** (1), 2–9.

Schaefer, R. B. (1999) Surface Discharge Lamp. U.S. Patent 5,945,790, Aug 31.

Schaefer, R. B. (2004) Surface Discharge Lamp and System. U.S. Patent 6,724,134 B1, April 20.

Scheible, O. K.; Casey, M. C.; Forndran, A. (1986) Ultraviolet Disinfection of Wastewaters from Secondary Effluent and Combined Sewer Overflows; EPA-600/2-86-005; U.S. Environmental Protection Agency: Cincinnati, Ohio.

Schulz, C.; Olejnik, D.; Feliers, C.; Cervantes, P.; Mysore, C. (2003) Full-Scale Testing of UV Irradiance Actinometry Sensors for Monitoring Dose Delivery in UV Reactors. *Proceedings of the International UV Association World Congress*; Vienna, Austria, July.

Setlow, R.; Carrier, W. L. (1966) Pyrimidine Dimers in Ultraviolet-Irradiated DNA's. *J. Molecular Biol.*, **17** (1), 237–254.

Setlow, J. K.; Setlow, R. B. (1967) Contribution of Dimmers Containing Cytosine to Ultra-Violet Inactivation of Transforming DNA. *Nature*, **213**, 907–909.

Shin, G.-A.; Linden, K. G.; Arrowood, M.; Sobsey, M. D. (2001) Low Pressure UV Inactivation and Subsequent DNA Repair Potential of *Cryptosporidium parvum* Oocysts. *Appl. Environ. Microbiol.*, **67** (7), 3029–3032.

Shin, G.-A.; Linden, K.G.; Faubert, G. (2009) Inactivation of *Giardia lamblia* Cysts by Polychromatic UV Emission. *Lett. Appl. Microbiol.*, **48** (6), 790–792.

Sinha, R. P.; Hader, D. P. (2002) UV-Induced DNA Damage and Repair: A Review. *Photochem. Photobiol. Sci.*, 1, 225–236.

Sommer, R.; Lhotsky, M.; Haider, T.; Cabaj, A. (2000) UV Inactivation, Liquid-Holding Recovery, and Photoreactivation of *Escherichia coli* O157 and Other Pathogenic *Escherichia coli* Strains in Water. *J. Food Protection*, **63** (8), 1015–1020.

Sosnin, E. A.; Oppenländer, T.; Tarasenko, F. V. (2006) Applications of Capacitive and Barrier Discharge Excilamps in Photoscience. *J. Photochem. Photobiol. C*, **7**, 145–163.

Strauss, J. H., Jr.; Sinsheimer, R. L. (1963) Purification and Properties of Bacteriophage MS2 and of Its Ribonucleic Acid. *J. Molecular Biol.*, **7** (1), 43–54.

Sutherland, J. C.; Griffin, K. P. (1981) Absorption Spectrum of DNA for Wavelengths Greater than 300 nm. *Radiation Res.*, **86** (3), 399–410.

Ta, N.; Hong, J.; Liu, T.; Sun, C. (2006) Degradation of Atrazine by Microwave-Assisted Electrodeless Discharge Mercury Lamp in Aqueous Solution. *J. Hazard. Mater.*, **138** (1), 187–194.

Tree, J. A.; Adams, M. R.; Lees, D. N. (1997) Virus Inactivation during Disinfection of Wastewater by Chlorination and UV Irradiation and the Efficacy of F+ Bacteriophage as a 'Viral Indicator'. *Water Sci. Technol.*, **35** (11-12), 227–232.

Tyagi, V. K.; Lo, S. L. (2013) Microwave Irradiation: A Sustainable Way for Sludge Treatment and Resource Recovery. *Renewable Sustainable Energy Rev.*, **18**, 288–305.

U.S. Environmental Protection Agency (2006) *Ultraviolet Disinfection Guidance Manual for the Final Long Term 2 Enhanced Surface Water Treatment Rule*; EPA-815/R-06-007; U.S. Environmental Protection Agency: Washington, D.C.

Verhoeven, S.; Bemus, R.; Bokermann, C.; Odegaard, C.; Jessica Patterson, J. (2013) *Bacillus pumilus* as a High Resistance Surrogate for High Dose UV Reactor Validation. *Proceedings of the International UV Association World Congress*; Las Vegas, Nevada, Sept 22–25; International Ultraviolet Association: Washington, D.C.

Von Sonntag, C.; Schuchmann, H. (1992) UV Disinfection of Drinking Water and By-Product Formation—Some Basic Considerations. *J. Water SRT-Aqua*, **41** (2), 67–74.

Wang, T.; MacGregor, S. J.; Anderson, J. G.; Woolsey, G. A. (2005) Pulsed Ultra-Violet Inactivation Spectrum of *Escherichia coli*. *Water Res.*, **39** (13), 2921–2925.

Wardle, B. (2009) *Principles and Applications of Photochemistry*; Wiley & Sons: New York.

Ware, M. W.; Augustine, S. A.; Erisman, D. O.; See, M. J.; Wymer, L.; Hayes, S. L.; Dubey, J. P.; Villegas, E. N. (2010) Determining UV Inactivation of *Toxoplasma gondii* Oocysts by Using Cell Culture and a Mouse Bioassay. *Appl. Environ. Microbiol.*, **76**, 5140–5147.

Whitby, G. E.; Palmateer, G.; Cook, W. G.; Maarschalkerweerd, J.; Huber, D.; Flood, K. (1984) Ultraviolet Disinfection of Secondary Effluent. *J. Water Pollut. Control Fed.*, **56** (7), 844–850.

Whitby, G. E.; Palmateer, G.; Cook, W. G.; Boon, F.; Janzen, E. (1985) The Effects of Wastewater Quality on Ultraviolet Light Disinfection. *Proceedings of Technology Transfer Conference No. 6, Part 2 Water Quality Research*; Ontario Ministry of the Environment: Toronto, Ontario, Canada.

Wong, M.; Kumar, L.; Jenkins, T. M.; Xagoraraki, I.; Phanikumar, M. S.; Rose, J. B. (2009) Evaluation of Public Health Risks at Recreational Beaches in Lake Michigan via Detection of Enteric Viruses and a Human-Specific Bacteriological Marker. *Water Res.*, **43** (4), 1137–1149.

Yoon, J. H.; Lee, C. S.; O'Connor, T. R.; Yasui, A.; Pfeifer, G. P. (2000) The DNA Damage Spectrum Produced by Simulated Sunlight. *J. Mol. Biol.*, **299** (3), 681–693.

Zhang, X. W.; Li, G. T.; Wang, Y. Z.; Qu, J. H. (2006) Microwave Electrodeless Lamp Photolytic Degradation of Acid Orange 7. *J. Photochem. Photobiol. A: Chem.*, **184** (1), 26–33.

Zimmer, J. L.; Slawson, R. M. (2002) Potential Repair of *Escherichia coli* DNA Following Exposure to UV Radiation from Both Medium- and Low-Pressure UV Sources Used in Drinking Water Treatment. *Appl. Environ. Microbiol.*, **68** (7), 3293–3299.

3

Bioassay Methods to Determine the Ultraviolet Dose (Fluence) Delivery of an Ultraviolet System

G. Elliott Whitby, Ph.D., and Bill Sotirakos

1.0 INTRODUCTION	52	
2.0 BIOASSAY PROTOCOL FOR WASTEWATER	53	
2.1 Planning and Preparation	54	
2.1.1 Test Ultraviolet System Characteristics	54	
2.1.2 Challenge Microorganisms Used in Validations	55	
2.1.3 Water Source Key Characteristics	57	
2.1.4 Absorbing Chemical	57	
2.1.5 Mixing and Sampling	58	
2.1.6 Lamp Variability and Ultraviolet Sensor Port Window Testing	58	
2.1.7 Measurement Equipment	58	
2.2 Inlet/Outlet Structures	58	
2.3 Test Lamps	59	
2.4 Test Conditions and Quality Assurance/Quality Control Samples	59	
2.5 Third-Party Oversight	60	
3.0 MICROBIOLOGICAL TESTING	61	
3.1 Preparing the Challenge Microorganism	61	
3.2 Verifying Ultraviolet Reactor Properties and Ultraviolet-Intensity Sensor Performance	61	
3.3 Measuring Ultraviolet Dose (Fluence) Delivery	61	
3.4 Collimated Beam Testing	62	
3.5 Validation and Data Analysis	63	
4.0 EXISTING DATA	63	
5.0 REPORTING	64	
6.0 BIOASSAY VALIDATION EXEMPLAR	65	
6.1 Validation Study	65	
6.2 Hydraulic Characterization	66	
6.3 System Parameters	67	
6.3.1 Power Measurements	67	
6.3.2 Ultraviolet Sensor Readings	67	
6.3.3 Headloss and Water Level	69	

6.4 Bioassay Testing	70	6.4.4 Bioassay Results and Data Analysis	74
6.4.1 Collimated Beam Analysis	71		
6.4.2 Bioassay Test Procedure	72	6.5 SUMMARY	76
6.4.3 Log Inactivation (Log I) Equation	73	7.0 REFERENCES	76

1.0 INTRODUCTION

To ensure that the treatment objectives of a UV disinfection system are met, it is important to validate, or verify, that the UV equipment is delivering the required UV dose (fluence) for the application. Determining the delivered UV dose (fluence) of a specific reactor is a means of overcoming issues associated with reactor hydraulics and also provides a way of comparing one reactor to another with respect to a delivered dose. The most widely accepted method for determining the UV dose (fluence) delivery performance is biodosimetry. Biodosimetry is a microbiological method that establishes the dose (fluence) for a UV system under specific conditions. While several protocols exist for completing biodosimetry tests, or bioassays, for different applications, only three methods are in wide-scale use for wastewater applications. These are as follows:

- *Ultraviolet Disinfection Guidelines for Drinking Water and Water Reuse*, 3rd edition, published by the National Water Research Institute (NWRI) in collaboration with the Water Research Foundation (NWRI and WRF, 2003, 2012), specifically, "Chapter 2: Water Reuse" and "Chapter 3: Protocols" (hereafter referred to as the "NWRI/WRF guidelines");
- *Ultraviolet Disinfection Guidance Manual for the Final Long Term 2 Enhanced Surface Water Treatment Rule* published by the U.S. Environmental Protection Agency (U.S. EPA) (2006) (hereafter referred to as "U.S. EPA *Ultraviolet Disinfection Guidance Manual*" or "U.S. EPA UVDGM"); and
- Uniform Protocol for Wastewater UV Validation Applications (Whitby et al., 2011) by the International Ultraviolet Association (IUVA) (hereafter referred to as the "IUVA protocol").

All three guidelines follow similar formats (Table 3.1) and are widely used by UV manufacturers, engineering consultants, and regulators.

TABLE 3.1 Format and equivalency of the three key UV validation protocols commonly used for wastewater applications.

U.S. EPA UVDGM (2006)	NWRI/WRF guidelines (2003, 2012)	IUVA protocol (2011)
Planning and preparation	Introduction	Planning and preparation
	Test facilities requirements and setup	
Testing	Microbiological testing	Microbiological testing
	Testing and sampling requirements	Testing and sampling requirements
Validation data analysis	Data analysis and reporting	Validation data analysis
Compare reactor hydraulics using CFD		Other methods
Reporting		Reporting

However, only the IUVA protocol specifically references the particular challenges associated with completing bioassays in wastewater applications. For the purpose of this chapter, the term wastewater applications refers to a biological treatment plant that is achieving an average effluent quality of less than 30 mg/L biochemical oxygen demand/total suspended solids and a disinfection permit requirement for a surface water discharge. This chapter will detail information that is required to perform a bioassay to determine the UV dose (fluence) delivery of a UV system. It will also briefly describe how to use this information to size a UV system for a water resource recovery facility (WRRF) and reference the more detailed chapters where sizing is described.

There are new and emerging technologies to aid in validation of UV reactors. Computational fluid dynamics (CFD) and Lagrangian actinometry with dyed microspheres are two such technologies. Information on these techniques is presented in Chapter 4 that describe supplemental analysis tools to provide greater flexibility and understanding of reactor design.

2.0 BIOASSAY PROTOCOL FOR WASTEWATER

The IUVA protocol format is similar to the U.S. EPA UVDGM and includes the following components:

- Planning and preparation,
- Microbiological testing,

- Validation data analysis,
- Additional analysis using advanced tools and existing data, and
- Reporting.

The treatment objectives and the UV dose (fluence) that is required at each particular WRRF must be determined before this bioassay method can be used to size the UV equipment. This may be accomplished through long-term measurements of flow, UV transmittance (UVT), and total suspended solids; additional information on these parameters is provided in Chapter 5. The UV dose (fluence) requirements may also be determined by preparing UV dose- (fluence)-response curves of the indicator microorganisms with a collimated beam apparatus to determine the required UV dose (fluence).

2.1 Planning and Preparation

It is important to understand the goals of testing, how testing will be completed, and within what limitations. This section describes key elements of planning and preparation for validation testing. Details of the test plan and final validation report must be reviewed by a third party so they conform to the bioassay protocol or regulatory requirements. Validation testing for wastewater applications can be conducted at a WRRF or a dedicated testing facility (manufacturer-owned, or third party). It is important to recognize that a bioassay validation is not a pass or fail test, rather, a means to determine the UV dose (fluence) delivery of a specific UV system. A summary of what information a bioassay is capable or not capable of providing is included in Table 3.2.

2.1.1 Test Ultraviolet System Characteristics

A typical wastewater bioassay test stand equipment configuration should include the following:

- The test unit must be equivalent in configuration and operation to the commercial unit in terms of components, including lamps, ballasts, quartz sleeves, sensors, control systems, and automatic cleaning devices and other fixed or moving devices such as baffles and support bars;
- The test unit must be hydraulically scalable or a commercially available full-scale module as per the NWRI/WRF guidelines. However, it is recognized that additional analysis using field measurements and/or advanced tools described in Chapter 4 could be used to

TABLE 3.2 Information that a bioassay of a UV system is capable or not capable of providing.

Information measured by bioassay	Information not measured by bioassay
Effect of flow	Effect of suspended solids
Effect of UV transmittance	Quartz sleeve fouling exceeding tested values
Effect of quartz sleeve fouling	Random lamps not producing UV light
Effect of end-of-lamp life	Flows above or below what has been tested
Effect of different organisms with different UV sensitivities on UV dose	Ultraviolet transmittances above or below what has been tested
Effect of UV lamp output turndown	Lamp output below test values

justify operational variations. Closed-vessel UV systems may not be hydraulically scaled; and

- For open-channel systems, a minimum of two banks in series shall be required because hydraulics and water level have a known effect on the performance of open-channel UV systems. For this reason, performance of UV banks may not be additive. Instead, performance of additional banks is a function of flow/lamp and water level. Full-scale UV systems often consist of multiple banks; accordingly, this protocol recommends consideration for standby capacity.

2.1.2 Challenge Microorganisms Used in Validations

Biodosimetry uses a challenge microorganism that is spiked into the UV system to determine the reduction equivalent dose (RED). Criteria for selection of a bioassay test organism have been described by a number of researchers (e.g., Sommer and Cabaj [1993]); for bioassay validation, it is critical that the challenge microorganism possess the following properties:

- Nonpathogenic to humans;
- Easy to grow in high concentrations and simple to enumerate;
- Demonstrates repeatable results and stability over long periods of time;
- Has a known action spectrum that correlates to the target organism;
- Is predefined to ensure valid comparisons between bioassays;
- Is analyzed only within the linear region of the UV dose- (fluence)- response curve;
- Must have UV inactivation kinetics that are similar to the indicator organisms or pathogens or two organisms must be used that span the

UV inactivation kinetics of the indicator organisms or pathogens, as described in Chapter 2; and
- Must not be indigenous because UV inactivation kinetics vary from site to site.

Based on the aforementioned criteria, there are several challenge microorganisms (or surrogates) that can be used; typical challenge microorganisms that have been used for bioassays include *Bacillus subtilis* and *Aspergillus* spores, MS2 coliphage, and other phage such as T1, T1UV, T7, PhiX184, and Qbeta. The use of a specific organism should correlate with the sensitivity of the pathogen target over the dose range of interest that is being verified. For example, MS2 can be used for dose (fluence) levels up to 120 mJ/cm^2, whereas T phage are typically used for dose (fluence) levels up to 25 mJ/cm^2. The dose (fluence) required for 2 log^{10} inactivation for common challenge microbes is defined in Table 3.3. Regardless of which surrogates are used, they must be prepared and used in accordance with the NWRI/WRF guidelines and U.S. EPA UVDGM.

For wastewater applications, it is preferable to use two surrogates that span the UV inactivation kinetics of the target (permit compliance) organisms. This allows UV systems to be sized on the dose-per-log (D_L) approach, where the target organism and surrogate should have a similar D_L of inactivation. The D_L approach requires a third-party validated bioassay to be performed with two challenge organisms that span the UV inactivation kinetics of the target organism. From the bioassay data, a dose equation based on D_L is developed. With the bioassay equation, and a collimated beam analysis for a target organism at a particular WRRF, the size of the UV system to comply with the permit limit or dose requirement can be established.

TABLE 3.3 Ultraviolet dose (fluence) to achieve 2-log (99%) reduction of select challenge microorganisms for UV disinfection systems (U.S. EPA, 2006). (UV dose [fluence] levels are in mJ/cm^2.)

Microorganism	Ultraviolet dose for 2 log (99%) reduction
Bacillus subtilis	39
MS2 phage	34
Qbeta phage	22.5
PhiX174 phage	5.3
E. coli	4.8
T7 phage	7.5

2.1.3 Water Source Key Characteristics

Key characteristics of source water for conducting bioassay testing should include the following:

- Finished water supply (potable water) and/or filtered (cloth or granular) WRRF effluent (secondary or tertiary), or a blend of both;
- Turbidity <2 nephelometric turbidity units or <5 mg/L total suspended solids in all instances; because suspended solids are different at each WRRF, a filtered effluent or potable water should be used;
- Must not contain disinfectant residuals that could affect the test microorganisms;
- The water must be dechlorinated (e.g., with thiosulfate or bisulfite) before being used for the bioassay; residual chlorine should be "nondetect", with a 0.05-mg/L detection limit. Quenching agents should not affect UVT in the 200- to 300-nm range and should not affect the challenge organism;
- The pH after UVT and dechlorination adjustment should be within ±0.5 pH units of the initial pH; otherwise, buffering is required; and
- The effect of additives on polychromatic absorbance shall be measured and documented. It is recommended not to use additives that may have an adverse effect on the polychromatic absorbance.

2.1.4 Absorbing Chemical

Where a UV-absorbing chemical is used to simulate the range of UVT values defined in the test plan, it is critical that it possess the following properties:

- An absorbance spectrum similar to the background filtered effluent/water used for validation. A full spectral scan in the UVC range is required if polychromatic lamps are used;
- Known and uniform effect on all relevant parameters;
- Not be toxic to the surrogate (challenge organism);
- Known spectral absorbance in the UVC wavelength range; and
- Solubility should not be affected by lamp heat dissipation.

Thus, the following UV-absorbing compounds are permissible: coffee, lignin sulfonic acid, or humic acids such as Super Hume™. Material safety data sheets for the absorbing compounds should be included in the validation report.

2.1.5 Mixing and Sampling

It is critical that any UV-absorbing chemical or challenge microorganism injected to the flow stream be uniformly dispersed at both the influent and effluent sampling points. Verification of mixing should be completed according to the method described in the U.S. EPA UVDGM and should be fully documented in the final validation report. The following guidance is provided regarding mixing and sampling for validation testing:

- The effluent sample point must be far enough downstream from the reactor exit such that fluid streamlines exiting the reactor have the opportunity to fully mix and disperse with each other to provide samples representative of post-reactor effluent;
- Inline mixing, with one mixer before the influent sample point and one mixer before the effluent sample point, is required for closed-vessel systems. The same is preferred for open-channel systems; however, if a mixer is only used before the bank of UV lamps, samples must be collected after the level control device; and
- The effluent sample location, particularly for open-channel systems, should eliminate free surface and wall edge effects.

2.1.6 Lamp Variability and Ultraviolet Sensor Port Window Testing

To account for lamp variability, the UV system supplier must include the NWRI and WRF test results for end-of-lamp life testing (NWRI and WRF, 2003). Only lamps that have been running inside of quartz sleeves under water and tested while operating under water will be acceptable. Air testing is not acceptable. Lamps that have not been tested or operated using this method should not be accepted. Closed-vessel systems should follow the U.S. EPA UVDGM procedures. Ultraviolet sensor port window testing should also follow the U.S. EPA UVDGM.

2.1.7 Measurement Equipment

All key process parameters are to be monitored and recorded, including flow, UVT, electrical power consumption, power input to the lamps (power to the lamps and ballasts may be substituted), UV intensity, water temperature, pressure (for enclosed vessels), and headloss. The methods described in the U.S. EPA UVDGM are comprehensive and should be adopted.

2.2 Inlet/Outlet Structures

Configuration of the inlet and outlet conditions must be documented in the validation report per the U.S. EPA UVDGM for closed-vessel reactors and per NWRI/WRF guidelines for open-channel systems. If the site-specific installation is different than the configuration used during validation testing, then the velocity profile or, preferably, the UV dose (fluence) and/or UV dose (fluence) distribution should be shown to be equivalent or better than that observed for the bioassay validation. This should be completed by one or more of the following methods, as appropriate: velocity profile as described in the NWRI/WRF guidelines (NWRI and WRF, 2003), CFD per Chapter 5 of this document, or check-point bioassay (NWRI and WRF, 2012).

2.3   Test Lamps

For wastewater applications, UV lamps should be documented to have been burned-in under water for at least 100 hours. Power to the UV lamps and UVT parameters should be adjusted to account for lamp aging and quartz sleeve fouling. Equipment verification for these two variables is briefly discussed in Section 4.0 of this chapter.

2.4   Test Conditions and Quality Assurance/Quality Control Samples

Validation test conditions should reflect as many variables as possible with respect to the wastewater and UV equipment. Some of these are described in the NWRI/WRF guidelines and U.S. EPA UVDGM. Therefore, the test matrix should be designed to a specified range of the following water qualities as defined by UVT, flows, and lamp output power levels regardless of the ultimate operating philosophy: UV dose (fluence) pacing, intensity pacing, or confirming existing validation equations (checkpoint bioassay). Adjusting the lamp output power to simulate end-of-lamp life is thought to be more realistic than using equations to adjust for a bioassay that is only performed with a new lamp at full power.

The challenge microorganism(s) should be injected into the flow upstream of the UV reactor under steady-state conditions. Good sampling practices should be followed to collect at least three influent and three effluent samples for each test condition. The following process measurements should be taken for each sampling event:

- Flow (volume per time),
- Ultraviolet intensity from all sensors (milliwatts per centimeter squared),
- Calculated UV dose (fluence) (millijoules per centimeter squared),
- Percent lamp current or lamp power (Watts),
- Ultraviolet transmittance,

- Electrical power (Watts), and
- Water temperature (Celsius).

Where any of the aforementioned parameters change in value by more than the error of measurement (see Section 5.5 of U.S. EPA UVDGM for error of measurement), over the course of each test condition, the test should be repeated. Standard quality control samples should be taken as follows:

- Reactor controls—with lamps off; log inactivation equivalent to RED <3% of lowest RED tested;
- Reactor blanks—without surrogate addition take one sample per day; the measured densities should be negligible;
- Trip controls—one sample bottle of the stock solution of the challenge microorganism should travel with the stock solution used for validation testing from the microbiological laboratory to the location of reactor testing and back to the laboratory. The change in log concentration of the challenge microorganism in the trip control should be within the measurement error; this is typically on the order of 3 to 5% (e.g., at 1×10^{11}, log = 11. ± 5%, equivalent to ~10.5 to 11.5);
- Method blanks—typical laboratory blanks are generated by measuring the microorganism in reagent water; and
- Stability blanks—the purpose is to show that no degradation of the surrogate occurred with time in the water containing the UVT modifier and in the water matrix.

2.5 Third-Party Oversight

An independent, third party must provide oversight of the validation study to ensure that testing and data analyses are conducted in a technically sound manner and without bias. Individuals qualified for such oversight include engineers experienced in testing and evaluating UV reactors and scientists experienced in the microbial aspects of biodosimetry. Appropriate individuals should have no real or apparent conflicts of interest regarding the ultimate use of the UV reactor being tested. A qualified third party should be present for, and direct all testing, analyze data, and author a detailed report. The final report should include the names and qualifications of all persons involved in the testing and their role. When appropriate, the third party should rely on additional outside experts to review various aspects of UV validation testing, such as lamp physics, optics, hydraulics, microbiology, and electronics.

3.0 MICROBIOLOGICAL TESTING

The U.S. EPA UVDGM contains a more comprehensive microbiological testing protocol than the NWRI/WRF guidelines and reflects the latest understanding of UV disinfection technology; therefore, this guideline should be adopted in its entirety for uniform validation testing for wastewater applications. However, specific unique challenges apply with wastewater and, therefore, the issues outlined in the following subsections should be considered.

3.1 Preparing the Challenge Microorganism

Stability should be checked and consistent recovery from seeded effluent should be confirmed, particularly in a treated wastewater effluent matrix. The challenge microorganism concentrations should be stable over the holding time between sampling and completion of the assays. If they are not stable, data will be unusable because distinguishing between the sources of inactivation, such as exposure to UV light and die-off in holding, will be impossible. Stability verification can help ensure that the bioassay and challenge microorganism samples will be viable and that the data will be useable. Generally, less-resistant organisms such bacteriophages T1, T7, and Q Beta are used to simulate indicator bacteria that are not protected by the suspended solids; this is the initial straight line portion of the UV dose- (fluence)-response curve. The MS2 coliphage is used to simulate the flat portion of the UV dose- (fluence)-response curve, where the indicator bacteria are protected by the suspended solids.

3.2 Verifying Ultraviolet Reactor Properties and Ultraviolet-Intensity Sensor Performance

Water temperature must be measured during the bioassay. Because water temperature cannot be varied during testing, the UV manufacturer must submit UV-intensity testing by a third party of the same lamp, ballast, and quartz sleeve combination at water temperatures from 5 to 30 °C. For medium-pressure systems, a temperature sensor and safety cutoff switch to prevent overheating should be provided by the manufacturer. Higher variability in sensor readings should be permitted if additional operational safety factors (i.e., setpoints) are included.

3.3 Measuring Ultraviolet Dose (Fluence) Delivery

To determine the UV dose delivered, calculations are based on data collected during the bioassay experiments; to provide good quality data for these calculations, the following conditions should be met:

- Ultraviolet transmittance should be within ±2% of the target UVT;
- Water temperature variability should be within 0.5 °C;
- Sampling shall not proceed until a minimum of five total void volume exchanges have passed through the UV system. This flush volume is calculated between the microorganism's injection point and the effluent sampling point;
- At least three influent and three effluent samples for each test condition should be collected. Influent and effluent samples are not collected at the same time, rather, they are collected in an alternating sequence at times that approximate the time of travel across the system. There should be at least one volume exchange between samples;
- Influent samples must be taken from the batch tank, from the feed pipe, or from the channel. Effluent samples must be taken from the reactor outflow or the channel after the effluent weir or from a sample tap, which is representative of the entire outflow;
- The surrogated organism should be counted at a minimum of two dilutions, with at least two replicates per dilution;
- The following parameters should be measured and recorded: flow, UV intensity, online UVT, calculated UV dose (fluence) (both before and after the samples are collected), UVT of each influent sample (measured with a UV spectrophotometer), electrical power consumed by the lamps and/or ballasts, ambient air temperature, and water temperature for each test. The sensor should conform to ÖNORM M 5873-1 and shall measure only the germicidal portion of the light emitted by the UV lamps as measured at 254 nm;
- Samples for UVT should be collected separately and be measured within 24 hours; and
- Concentration of the challenge microorganisms, before and after exposure to UV light, should be measured within 24 hours of sample collection unless stability studies indicate samples are stable over longer periods of time. Samples should be stored in the dark at 4 °C or on ice immediately after being collected. Exposure of samples to visible light should be avoided.

3.4 Collimated Beam Testing

Protocols for collimated beam testing should follow those in the literature; additional information on collimated beam testing is provided in Chapter 2. The UV sensitivity of the challenge microorganism(s) and shape of each

UV dose- (fluence)-response curve should be consistent with the expected inactivation behavior; accordingly, confidence bands developed for MS2 and other surrogates should be used as a test of the quality of the UV dose- (fluence)-response data. In the case of a challenge microorganism with a shoulder or tailing in the UV dose- (fluence)-response curve, UV sensitivity should be defined as the sensitivity over the linear region of inactivation that occurs between the shoulder and the onset of tailing. The shoulder is a flat portion at the initial start of the microorganism's UV dose (fluence) response where it is resistant to UV light. The flat portion at the end of the linear portion of the UV dose- (fluence)-response curve is typically a result of the microorganisms being protected because of clumping, which prevents penetration of UV light into the center of the clumps. Organisms with a shoulder are not recommended for developing a bioassay.

3.5 Validation and Data Analysis

Experimental data should be documented, preferably in tabular format, and included in the validation report. The RED should be calculated for each experiment using a combination of reactor testing data and collimated beam results. Additional analysis of RED data depends on the reactor's UV dose- (fluence)-monitoring strategy. For the UV-intensity-control approach, RED results are averaged for each test condition and evaluated to identify the minimum value. For the calculated dose- (fluence)-control approach, all RED values and associated test conditions are used to create a UV dose- (fluence)-monitoring equation. Validation testing produces the following types of data for each experiment:

- Concentration of the challenge microorganism(s) in the influent and effluent samples,
- Ultraviolet transmittance of the water,
- Flowrate,
- Ultraviolet intensity as measured by the UV sensor,
- Lamp or lamp and ballast power, and
- Status (on/off) for each lamp.

4.0 EXISTING DATA

Ultraviolet equipment validated before the publication of the IUVA protocol (Whitby et al., 2011) should be recognized as long as the validation was conducted according to the following factors:

- The microorganisms used in the previous validation were the same as those recommended in this publication;
- Quality assurance/quality control procedures that are generally inline with this document were followed;
- Data analysis was generally inline with the methods outlined in this document; and
- A qualified third party conducted and certified the results of the bioassay.

It should also be recognized that, in addition to UV dose (fluence) delivery performance validations, there are other related equipment tests that are used to verify operational performance. These include

- Lamp output measurement,
- Lamp age factor testing, and
- Cleaning mechanism (quartz sleeve fouling) testing.

While these verification tests can be completed separately to bioassay testing, it is recognized that they are used in the final equipment sizing design and, as such, deserve attention. The IUVA Manufacturers Council has published a protocol for the measurement of the UV output of low-pressure lamps, and it is recommended that this be used for measuring the UV output of lamps in air. A newer protocol will be published in 2015 in *IUVA News* that eliminates problems that were observed in the original method. A similar protocol for medium-pressure lamps is pending from IUVA. Separate, updated protocols for lamp aging and quartz fouling are required; however, in the short term, it is recommended that the existing NWRI/WRF guideline is followed (NWRI and WRF, 2003).

5.0 REPORTING

A formal validation report is an important element of any validation testing. Both the NWRI/WRF guidelines and the U.S. EPA UVDGM include reporting guidelines. Because the U.S. EPA UVDGM details a more comprehensive outline of the key elements of a validation report, together with checklists helpful for review and approval, this guideline should be used for the wastewater UV validation protocol. It is important that the validation report include all of the details of the testing and that the final bioassay is reported along with the validation envelope (i.e., range of test conditions and results) that it can be used within.

6.0 BIOASSAY VALIDATION EXEMPLAR

The bioassay should incorporate at least two surrogates with different UV sensitivities so they span the UV inactivation kinetics of the target organism/s in the disinfection permit. This approach is known as the dose-per-log inactivation (or D_L) method, where the bioassay equation can be applied to native wastewater organisms that are bracketed by surrogate organism sensitivity. The validation protocol exemplar presented utilizes T1 and MS2 coliphage as indicator organisms, with D_L values of 5 and 20 mJ/cm²/log inactivation, respectively. This method allows UV systems to be sized for permit compliance because the RED can be selected based on known or measured UV sensitivity (D_L) of the target organism, eliminating the need to estimate how fecal coliform, *E. coli* or Enterococci RED equates to MS2 RED, where each organism has a different sensitivity. An example using this method for UV equipment sizing is provided in Chapter 5.

6.1 Validation Study

As a hypothetical example, ABC Consulting ("ABC") was contracted by XYZ Manufacturer ("XYZ") to perform validation testing on their XYZ UV disinfection system. The validation testing was performed at a WRRF in "Anywhere", U.S.A. ABC witnessed field testing and performed data analysis; XYZ performed the validation testing and operated the equipment. Testing was conducted in accordance with the IUVA protocol; test microorganisms included MS2 and T1. Bioassay testing was conducted by adding the concentrated nonpathogenic surrogate organisms to the influent water; samples were collected from the influent and effluent of the UV reactor to determine the inactivation of the organisms through the disinfection system for a range of flowrates, power settings, number of operating banks (1 and 2) in series, and water UVTs.

For this exemplar analysis, the UV disinfection system is an open-channel UV system with low-pressure, high-output amalgam lamps that are horizontal and parallel to the flow. The system consists of one channel with two UV banks; each bank is comprised of four UV modules, each with four UV lamps for a total of 32 lamps (Table 3.4). Future designs of this system may use several banks in series as long as headloss remains acceptable, dose additivity is proven, and the design is within the validated range. For this example, influent to the UV system was filtered effluent from the WRRF, with a UVT of 80%. Flow was controlled by a pump and valve upstream of the UV system and was measured using an electromagnetic flow meter. A static mixer, upstream of the UV system, was used for mixing microorganisms and the UVT modifier into the influent.

TABLE 3.4 Description of the UV system from Manufacturer XYZ that was tested with a bioassay.

System type	Open channel
Sleeve cleaning system*	Automated mechanical cleaning mechanism
Reactor lamp type	Low-pressure, high-output amalgam lamps
Lamp power	420 W with UVC-output at 160 W, measured using the IUVA Method for Measurement of Output of Monochromatic (254 nm) Low Pressure UV Lamps
Lamp part number	123456
Mounting position of lamp	Horizontal and parallel to the flow
Lamp spacing	133.4 mm (5.25 in.), centerline to centerline
Number of channel	1
Number of banks/channel	2
Number of UV modules/bank	4
Number of lamps/module	4
Total number of lamps	32
Operating approach	"Dose-pacing" method, relying on UV sensor readings
Sensor	One intensity sensor per bank
Ballast	SUPER-1 electronic ballast

*The cleaning mechanism was operated before each test run, but not during testing.

6.2 Hydraulic Characterization

The time required to achieve the steady-state concentrations of a dosed constituent in the UV reactor must be determined before testing. For this example, the time was determined by injecting a UVT modifier and sampling downstream at specific locations. Hydraulic testing was performed at 1.21 and 15.15 m^3/min (320 and 4003 gpm) (Dosing of constituents (i.e., UVT modifier and challenge organism) was performed with a dosing station that fed a sidestream of solutions into the pipe upstream of the static mixer before the UV channel. Influent samples were collected upstream of the first UV bank; effluent samples were collected just downstream of the manually controlled weir. Based on the results of hydraulic testing, the UV system was allowed to operate for 3 to 5 hydraulic residence times before sample collection as a measure of conservatism.

6.3 System Parameters

Water quality and operating parameters were recorded over the duration of this system validation and included power, UVT, flow rate, UV-intensity readings, and temperature. Concurrent measurement of online sensor readings, UVT, flow rate, ballast power, and headloss (upstream and downstream water level measurements) were performed for each experiment (test condition). The frequency of data collection is summarized as follows:

- Flow—before each sampling event,
- Ultraviolet transmittance—grab samples taken concurrent with each bioassay influent and effluent sample,
- Power—before each sampling event,
- Ultraviolet intensity—before each sampling event, and
- Temperature—measured during each test.

6.3.1 Power Measurements

Documentation of power draw and variability during validation allows for a better understanding of performance for full-scale systems. Comparison of ballast power settings and average online power readings for a single UV lamp is presented in Figure 3.1. There was limited variation in the power reading at each setting, thus measurements were taken for an entire bank of 16 lamps then divided by 16 to develop the power required on a per-lamp basis.

6.3.2 Ultraviolet Sensor Readings

The UV sensor is integral to the performance monitoring of this system. Sensor readings were taken during each test. The reference sensors were the same model and met the same criteria as the main duty sensor. Sensors are factory-calibrated, 254-nm specific sensors and do not need to be adjusted in the field. The sensors have excellent sensitivity for monitoring and control of UVC output at 254 nm. For this example, there was little variation among the intensity readings for each sensor. Sensor accuracy was assessed using four duty sensors and two reference sensors. The difference between duty sensors and reference sensors tested at 100% power averaged less than 3%. Ultraviolet sensor results are independent of the test organism; sensor data obtained during testing for both organisms were used.

A regression analysis was performed on the sensor data to develop an equation to predict sensor value (eq 3.1). Grubb's test can be used to determine

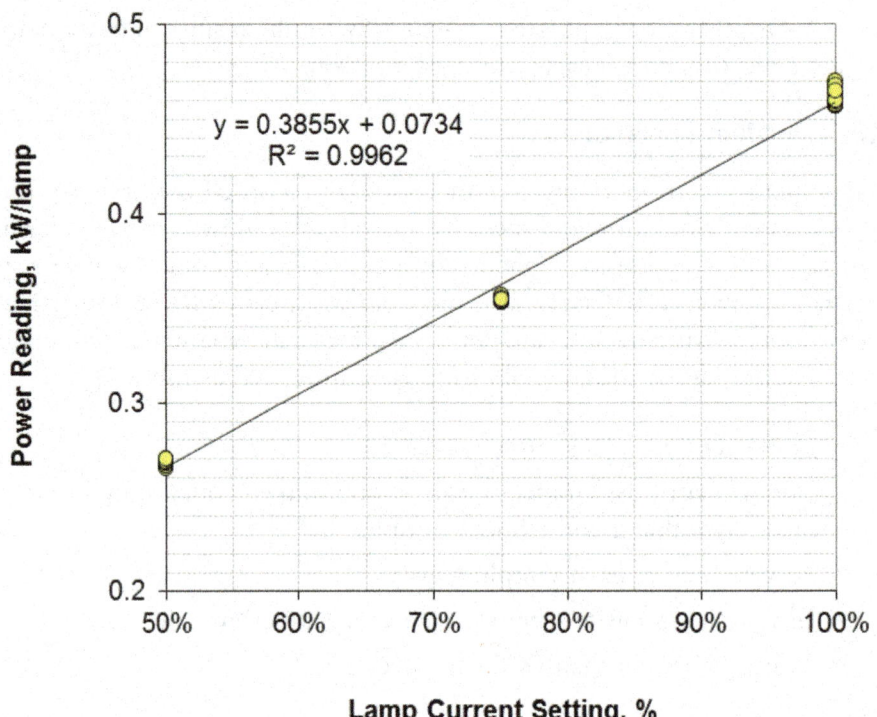

FIGURE 3.1 Power measurement per lamp vs lamp current setting.

outliers. For this example, the Grubb's test identified three data point outliers, which were removed from the analysis. The equation is as follows:

$$S_{pred} = 10^{-2.912} \times UVT^{4.2645} \times Power^{1.0562} \qquad (3.1)$$

Where
S_{pred} = predicted sensor intensity (mW/cm^2) (S_{pred} equals S_0 when the power to lamp is 100%),
UVT = percent UV transmittance (e.g., 65% = 65), and
Power = percent lamp current setting (e.g., 100% = 100).

The R^2 value for the predicted sensor value was 0.9899. Two MS2 tests and one T1 test were identified as outliers through Grubb's analysis and were removed from the sensor data set. All terms were significant at a 95% confidence level (p-values $\ll$ 0.05).

Instead of testing to determine lamp aging factors, lamp validation was based upon "sensor-based control", relying on accurate sensor readings to maintain dose, similar to drinking water UV applications. The aforementioned data that were collected and analyzed will be used with other parameters for development of a control equation.

6.3.3 Headloss and Water Level

Headloss testing was conducted over a range of flows from 1.21 to 15.14 m^3/min (320 to 4000 gpm). Water-level measurements are best measured by survey instruments, although hand measurements may suffice. During each test, depth of water before the first bank and after the last bank were measured. Headloss was calculated as the difference between these water depths divided by the total number of banks (2). Figure 3.2a shows headloss

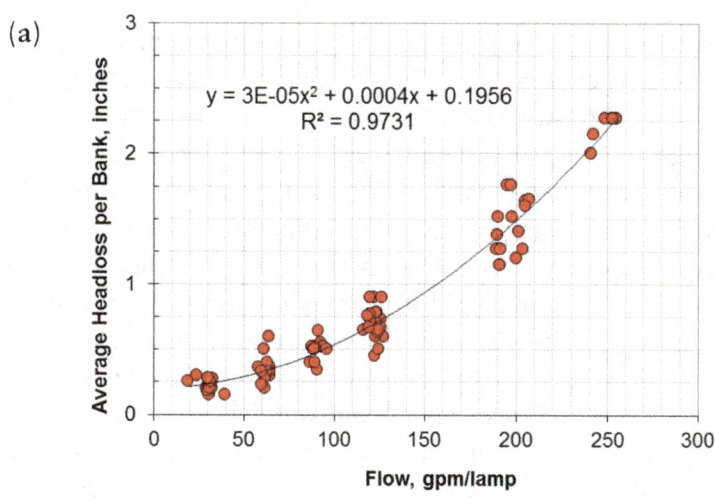

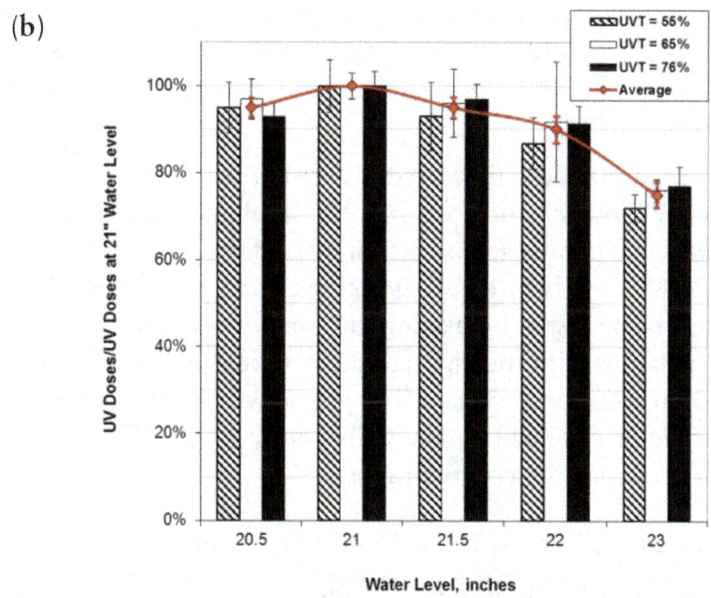

FIGURE 3.2 Headloss and water level data for bioassay exemplar. (a) System headloss as a function of flow and (b) system performance with varying water level and UVT headloss per bank vs flowrate.

as a function of flow, and the relationship was fitted using the following equation:

$$\text{Headloss} = 3 \times 10^{-5} \times Q^2 + 0.004 \times Q + 0.1956 \qquad (3.2)$$

Where
  Headloss = average headloss per bank, cm (in.), and
  Q = flow divided by the number of lamps in a single bank, L/m per lamp (gpm per lamp).

Tests were performed during the bioassay to determine the effect of water level on disinfection performance. While disinfection results are presented in later sections, results from tests at various water levels are presented here. Bioassay testing was completed with the effluent weir set so the target water level downstream of the tested bank was 53 cm (21 in.) 6.7 m (2.6 in.). above the centerline of the top lamp). Additional tests were conducted at specific depths above and below typical operation; these data were used to determine the acceptable range of water level for operation. Tests were performed at 100% lamp current (full power) and at constant target flows for each target UVT. The effect of water level on the system's performance is presented in Figure 3.2b. The UV module tested is a 4-lamp module; use of a larger module (e.g., 8-lamp module) may lessen the effect of water level because a proportionally smaller flow would flow over the top lamp. It is important to note that the water level data presented here, unless noted, were not included in the bioassay regression; additional analysis is presented later in this chapter.

6.4   Bioassay Testing

This section summarizes the measured UV disinfection performance for a range of flow, UVT, power, and bank number combinations and develops a performance/sizing equation for monitoring UV dose delivery. Lamps were burned-in for 100 hours before testing. It is necessary to test the UV output of all UV lamps being tested by placing the lamps next to the UV sensor in the reactor. The lamps with the highest output should be placed next to the UV sensors to provide conservatism. The protective quartz sleeves (around the lamps) should be cleaned by hand daily before testing. The following is a summary of the bioassay test conditions:

- Range of channel flows, 1.2 to 15 m³/min (320 to 4000 gpm);
- Range of UVT values, 40.2 to 77.8%;
- Range of sensor intensities, 0.52 to 18.60 mW/cm²;
- Ballast power operating range, 50 to 100%; and
- Tested range of flow per lamp, 0.08 to 0.9 m³/min (20 to 250 gpm) per lamp.

6.4.1 Collimated Beam Analysis

Collimated beam testing of the MS2 and T1 stock solutions must be performed during bioassay testing to verify that the stock used for the validation meets quality control requirements. The stock solution is acceptable if the MS2 UV dose response falls within the quality control ranges of the NWRI/WRF guidelines; the collimated beam protocol met the requirements of the U.S. EPA UVDGM. For this exemplar, the collimated beam data for MS2, bracketed by the quality control boundaries of the NWRI/WRF guidelines, are presented in Figure 3.3a; collimated beam data for T1

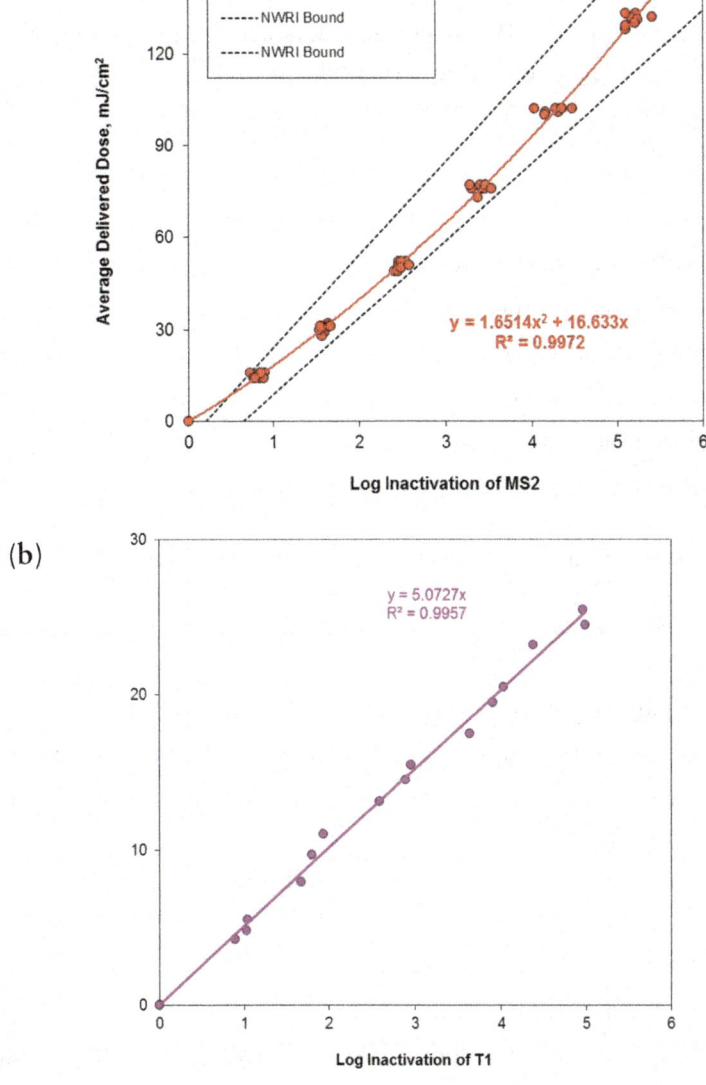

FIGURE 3.3 The (a) MS2 and (b) T1 collimated beam results, respectively.

are presented in Figure 3.3b. Analysis of MS2 data included the low-dose points (<20 mJ/cm^2), and the analysis approach provided in Appendix C of the U.S. EPA UVDGM was used for the estimation of the zero-dose point. Regression curves were fitted to the collimated beam data to develop the following UV dose-response relationships:

$$\text{UV Dose} = 1.6514 \times (\text{Log I})^2\ 16.633 \times \text{Log I} \quad \text{for MS2 only} \quad (3.3)$$

$$\text{UV Dose} = 5.0727 \times \text{Log I} \quad \text{for T1 only} \quad (3.4)$$

where Log I is the log inactivation value of MS2 or T1 coliphage.

6.4.2 Bioassay Test Procedure

Hypothetical testing of the UV system was performed to document the Log I applied by the system over a range of flowrates, UVT values, and power settings. The following parameters were concurrently monitored with each set of biological tests:

- Flowrate,
- Power draw (in kilowatts),
- Ultraviolet sensor values,
- Ultraviolet transmittance,
- Total and free chlorine,
- Water temperature, and
- Number of banks in service.

Biodosimetry was conducted using the following procedure:

1. Collect raw water sample and confirm absence of chlorine before bioassay testing begins;
2. Set the ballast power as required for the test condition;
3. Set the flow to the target value by adjusting the flow control valve;
4. Set the water UVT to the target value by adjusting the Super Hume™ solution pump;
5. Confirm accuracy of the UV spectrophotometer; re-zero with distilled water as necessary;
6. Confirm influent UVT value;
7. Initiate injection of coliphage;
8. Manually record flow, online UV intensity, online power, channel water level, and water temperature;

9. Collect three influent and three effluent samples for challenge microbe analysis;
10. As required, collect a sample from the influent sampling port that will be used to measure the UV dose response of the coliphage;
11. Collect effluent samples for measuring UVT either before or after microbial sampling and in parallel with microbial sampling; and
12. Manually record the UVT values.

Sample vials were prelabeled before testing according to the test plan. Documentation of the test sequence was recorded on data sheets. All microbiological samples were stored on ice until shipment by overnight courier to a laboratory for analysis of challenge organism concentrations. Three influent and three effluent samples were collected for each test from locations where water was completely mixed. All UVT samples were collected from the same influent and effluent sampling location as biological samples.

6.4.3 Log Inactivation (Log I) Equation

The Log I for the exemplar UV system is a function of flowrate (Q), UVT, and normalized UV sensor value (S/S_0). A good fit to validation data takes the following form:

$$\text{Log I} = (10^a \times A_{254}^A) - \left[\frac{S/S_0}{Q \times \text{Sensitivity}} \right]^{(c \times \text{Ln}(A_{254}) + d)} \times \text{Banks}^e \quad (3.5)$$

Where
- Log I = log inactivation values of challenge microorganisms (MS2 or T1);
- A_{254} = UV absorbance at 254 nm/cm;
- S = measured UV sensor value (mW/cm^2);
- S_0 = UV sensor value at 100% lamp power (new lamps) with clean sleeves, typically expressed as a function of UVT (mW/cm^2);
- Q = flow rate per lamp, calculated as the total flow in gpm divided by the number of lamps in one bank L/min (gpm) per lamp;
- Sensitivity = sensitivity of test organisms (MS2 or T1), calculated as the ratio of UV dose levels to Log I (RED/Log I);
- Banks = number of operating banks in series; and
- a, b, c, d, e = model coefficients obtained by fitting the equation to the data.

The UV dose algorithm used by the UV system controls includes an "S/S_0" term. This is a measurement of the relative lamp output that accounts for any degradation in lamp output compared to that expected for new lamps, operated at full power with clean quartz sleeves and sensor windows, as follows:

$$RLO = S/S_0 \qquad (3.6)$$

Where
RLO = relative lamp output,
S = measured UV sensor value (mW/cm^2), and
S_0 = UV intensity at full lamp power (new lamps) with clean sleeves (mW/cm^2).

6.4.4 Bioassay Results and Data Analysis

Testing was performed to determine flow-specific performance of the UV system for flows from 72 to 908 m^3/h (0.46 to 5.76 mgd) [(0.08 to 0.9 m^3/min [20 to 250 gpm] per lamp) at UVTs ranging from 40.2 to 77.8% and sensor intensities ranging from 0.52 to 18.6 mW/cm^2. Tests were performed with all 16 lamps per bank in operation and the water level was set at 53 cm (21 in.). Bioassay test results are used to determine UV system disinfection performance for MS2 and T1 inactivation. A statistical analysis performed on the bioassay data results in a performance equation that can be used for design and operation of a UV system for wastewater disinfection, meeting all of the specified parameters of this testing (i.e., flow rate, UVT, and UV-sensor intensity).

Data variations are captured in linear regression analysis; the regressed equation has a calculated R^2 value that ranges from 0 to 1, indicating how much of the variation is described by the equation. For example, an R^2 value of 0.90 indicates that 90% of the variation in the data is accounted for by the resulting equation. After performing a multiple linear regression on the data set as described previously, all three parameters were significant (i.e., their "p-values" were <0.05). A p-value of >0.05 indicates that an independent variable is not significant to the determined equation. The resultant equation for calculating Log I as a function of the operating conditions is as follows:

$$\text{Log I} = (10^{0.1612} \times A_{254}^{-4.188}) \times \left[\frac{S/S_0}{Q \times \text{Sensitivity}} \right]^{(-0.1528 \times \text{Ln}(A_{254}) + 0.6669)} \times \text{Banks}^{0.9454} \qquad (3.7)$$

$$\text{Sensitivity}_{\text{MS2}} = 1.6514 \times \text{Log I} + 16.1865 \qquad \text{for MS2 only} \qquad (3.8)$$

$$\text{Sensitivity}_{\text{T1}} = 5.0727 \qquad \text{for T1 only} \qquad (3.9)$$

Where
- Log I = log reduction values of MS2 or T1;
- A_{254} = UV absorbance at 254 nm/cm = $-\log$ (UVT/100);
- S = measured UV sensor value (mW/cm^2);
- S_0 = UV intensity calculated from eq 3.1 at full lamp power (0.459 kW/Lamp) (mW/cm^2);
- Q = flow rate per lamp, calculated as L/min (gpm) divided by the number of lamps in one bank;
- Sensitivity = sensitivity of test coliphage, calculated as the ratio of UV dose levels and Log I (RED/Log I); and
- Banks = number of operating banks.

A plot of predicted vs measured Log I is shown in Figure 3.4. The fit of predicted vs measured Log I values yielded a slope of 0.9993, with an R^2 of 0.9461 and a p-value of 2.08×10^{-105} at a 95% confidence level, indicating that the equation is an excellent fit to the data. Log I can be converted to UV dose by using eqs 3.3 and 3.4.

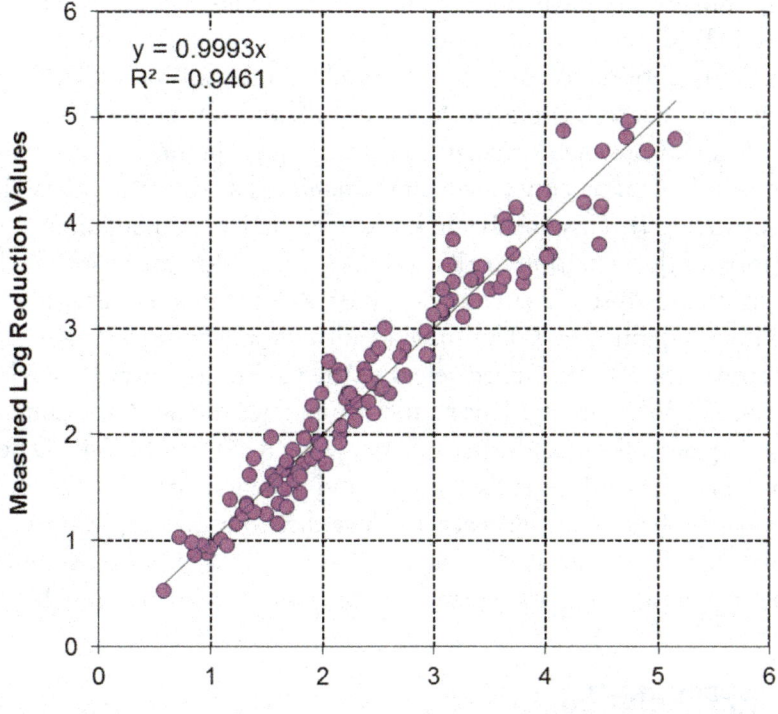

FIGURE 3.4 Comparison of predicted and measured log I values for the UV system from Manufacturer XYZ that was tested with a bioassay.

TABLE 3.5 Validation and operating range for use of the log I equation for the example UV disinfection system validation report.

Flow rate range[1]	UVT range[2]	Lamp power operating range	Sensor value operating range
≤0.954 m³/min/lamp (≤252 gpm/lamp)	≥40.2%	0.266–0.459 kW	0.52–18.64 mW/cm²

Notes:
(1) At flowrates below 0.076 m³/min (20 gpm) per lamp, this value must be used as the default value in the RED calculation.
(2) At UVT values above 77.8%, this value (77.8% UVT, or 0.109/cm A_{254}) should be used as the default value in the RED calculation.

6.5 SUMMARY

For this exemplar, a total of 88 bioassay samples were conducted. The MS2 and T1 dose-response curves generated by the collimated beam analysis were used to convert the measured log reduction values for each test into the measured UV dose values. A regression analysis was performed on the data to develop equations to calculate the predicted sensor value (eq 3.1), predicted log inactivation (Log I) values (eq 3.7), and UV dose values for MS2 and T1 (eqs 3.3 and 3.4, respectively). The Log I equation incorporates the challenge organism sensitivity, which is calculated as the ratio of UV dose to Log I (dose/Log I). This approach, known as the D_L method, allows for the application of this equation to all native organisms that are bracketed by the indicator organism sensitivity; for this bioassay, organism sensitivity ranged from 5 to 20 mJ/cm²/Log I. This method allows for accurate sizing of UV systems because the RED specified refers to the native organism with a known or measured UV sensitivity. This method eliminates the need or estimation of trying to equate a fecal coliform or *Enterococci* RED to a MS2 RED because each organism has a different sensitivity and its inactivation through the UV reactor differs based on the reactor's dose distribution. Application of this approach for UV sizing is appropriate within the conditions summarized in Table 3.5, and an example using this bioassay is provided in Chapter 5.

7.0 REFERENCES

National Water Research Institute; Water Research Foundation (2003) *Ultraviolet Disinfection Guidelines for Drinking Water and Water Reuse*, 2nd ed.; National Water Research Institute: Fountain Valley, California.

National Water Research Institute; Water Research Foundation (2012) *Ultraviolet Disinfection Guidelines for Drinking Water and Water Reuse*, 3rd ed.; National Water Research Institute: Fountain Valley, California.

Sommer, R.; Cabaj, A. (1993) Evaluation of the Efficiency of a UV Plant for Drinking Water Disinfection. *Water Sci. Technol.*, **27** (3-4), 357–362.

U.S. Environmental Protection Agency (2006) *Ultraviolet Disinfection Guidance Manual for the Final Long Term 2 Enhanced Surface Water Treatment Rule*; EPA-815/R-06-007; U.S. Environmental Protection Agency: Washington, D.C.

Whitby, G. E.; Lawal, O.; Ropic, P.; Shmia, S.; Ferran, B.; Dussert, B. (2011) Uniform Protocol for Wastewater UV Validation Applications. *IUVA News*, **13** (2), 26–33.

4

Innovations and Advances in Ultraviolet Reactor Analysis and Validation

Ernest (Chip) R. Blatchley III, Ph.D., P.E., BCEE, F. ASCE; Karl Scheible; and Chengyue Shen, Ph.D., P.E.

1.0 INTRODUCTION	79	4.0 EMERGING STRATEGY FOR REACTOR VALIDATION: STOCHASTIC APPROACH	91
2.0 FACTORS AFFECTING ULTRAVIOLET DISINFECTION REACTOR PERFORMANCE	81	5.0 REFERENCES	92
3.0 CONTEMPORARY METHODS FOR ULTRAVIOLET REACTOR VALIDATION	83		

1.0 INTRODUCTION

Protocols for design and validation of UV disinfection systems have been developed by many governmental and nongovernmental organizations. Table 4.1 provides a summary of the key features of common protocols. To varying degrees, all of the protocols listed in Table 4.1 involve empiricism and (necessary) conservative assumptions or factors of safety. However, none of these protocols explicitly or completely accounts for the fact that contemporary, continuous-flow UV reactors deliver a distribution of UV doses. Methods that fail to account for, and accurately quantify, the UV dose distribution delivered by a reactor cannot be used to directly define safe operating conditions for a reactor system. These methods must account for this lack of information through the use of engineering conservatism and generous safety factors.

TABLE 4.1 Summary of current guidance relative to UV disinfection design and validation.

Protocol name, sponsoring organization	UV-specific guidance	Other, related guidance issues
Ten States Standards (GLUMRB, 2004)	• Open-channel, modular units required • At least two banks in series required • UV dose ≥30 mJ/cm^2 • UV transmittance$_{254}$ ≥65%	• Biochemical oxygen demand ≤30 mg/L • Total suspended solids ≤30 mg/L • Hydraulic properties simulate plug flow • UV dose not defined

Ultraviolet Disinfection Guidelines for Drinking Water and Water Reuse (NWRI and WRF, 2012)	• Guidance for reuse applications:
	  >5 log$_{10}$ poliovirus inactivation
	  ≤2.2 most probable number/100 mL *total coliform*
	• RED ≥100 mJ/cm^2 preceded by media filtration (UVT$_{254}$ ≥55%)
	• RED ≥80 mJ/cm^2 preceded by microfiltration or ultrafiltration (UVT$_{254}$ ≥ 65%)
	• RED ≥50 mJ/cm^2 preceded by reverse osmosis (UV transmittance$_{254}$ ≥90%)
		• CFD not allowed
		• Spot-check biodosimetry used to validate full-scale systems
		• Definitions provided for *reduction equivalent dose, design UV dose,* and *operational UV dose*
		• Standardization of MS2 as challenge organism for biodosimetry

Ultraviolet Disinfection Guidance Manual for the Final Long Term 2 Enhanced Surface Water Treatment Rule Manual (U.S. EPA, 2006)	• Drinking water applications • RED by biodosimetry • Adjust for uncertainties by validation factor • Full-scale testing	• CFD allowed as supplementary tool to validation • Requirements on approaching piping configuration or hydraulics
Low Dose UV Systems for Secondary Effluent Applications (IUVA Manufacturers Council, 2012)	• Treated wastewaters • RED by biodosimetry • Scaled testing • Incorporate low-dose surrogate	• Corrects potential issues with using MS2 alone for low-dose applications • Recognizes the need to use validated systems when designing for wastewater applications
ÖNORM (2001)	• RED ≥40 mJ/cm^2 • *B. subtilis* spores used as challenge organism	
DVGW (2006)	• RED ≥40 mJ/cm^2 • *B. subtilis* spores used as challenge organism	

Systems that are designed and implemented following these approaches will generally comply with treatment requirements; however, these systems tend to be overdesigned and inefficient in terms of electrical power use.

Methods are available to accurately quantify or simulate the dose distribution delivered by a UV reactor. Application of these methods offers the potential to decrease capital and operating costs of these systems while improving process reliability. The purpose of this chapter is to present these methods and to describe how they may be applied to improve system designs.

2.0 FACTORS AFFECTING ULTRAVIOLET DISINFECTION REACTOR PERFORMANCE

In all photochemical processes, the dose of radiation delivered to the photochemical target is the master variable. The dose delivered to an individual photochemical target will determine the likelihood of a photochemically induced change taking place in that target. Within a population of targets, the delivered dose will determine the extent of photochemical conversion. In its most basic form, the dose of radiation may be defined as follows:

$$\text{Dose} = \int_0^\tau F(t) \cdot dt \tag{4.1}$$

Where
  Dose = dose delivered to a photochemical target $[=]$ mJ/cm^2,
  $F(t)$ = time-history of fluence rate delivered to photochemical target $[=]$ mW/cm^2,
  τ = period of exposure $[=]$ s, and
  t = time $[=]$ s.

Reactor performance in UV disinfection systems is quantified based on the fraction of target organisms that are inactivated (N/N_0) or the concentration of viable target organisms that remain in the treated water (N). Generally, reactor performance in UV disinfection systems will depend on the concentration of viable target organisms in the untreated water (N_0), the UV dose-response behavior (i.e., intrinsic kinetics) of the target organism, and the dose distribution delivered by the reactor to the target organism population (Figure 4.1). Well-defined methods are available to allow for quantification of each of these factors.

The concentration of viable (infective) target organisms is measured by application of standard microbiological methods. For bacteria, these methods often involve incubation in the presence of a substrate that is specific for the target bacterium. Viruses require incubation under conditions that allow the virus to complete its life cycle by infection of its host. Methods

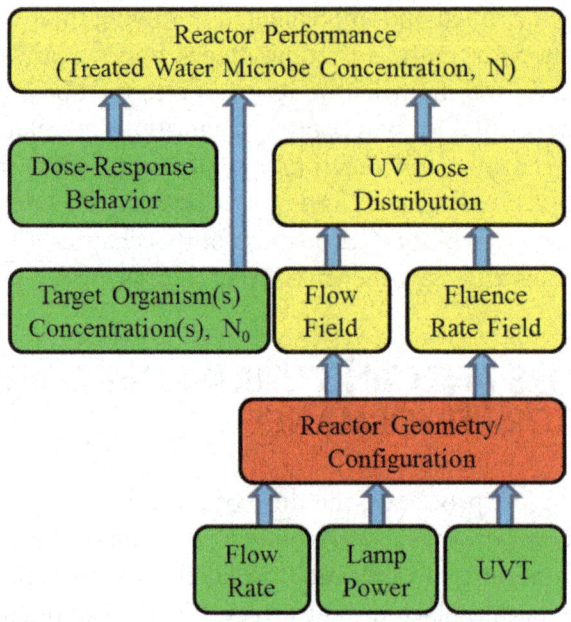

FIGURE 4.1 Schematic illustration of factors that affect performance in UV disinfection systems; arrows indicate direction of dependence among factors. The red box indicates fixed (unchanging) attributes of the system for a given reactor. Factors in green are variable; factors in yellow are variable, but depend on other attributes of the system. All factors are measurable by well-defined, standardized methods.

to measure infectivity of protozoa also typically involve the use of a host cell line or animal host.

Laboratory-based methods have also been developed for quantification of UV dose-response behavior. The most commonly applied of these methods involve the use of a collimated UV source and a shallow, well-mixed batch reactor (Bolton and Linden, 2003). In principle, the UV dose-response behavior of any organism can be quantified if an appropriate assay is available to quantify viability of the target organism.

As indicated in Figure 4.1, the dose distribution delivered by a UV reactor will be governed by fluid mechanics (i.e., flow field) and the fluence rate field. Accurate quantification of these two system attributes requires the use of methods that account for the three-dimensional characteristics of the flow and fluence rate fields. Generally, the flow field within a UV reactor will be governed by the flowrate of water imposed on the reactor and the physical boundaries of the fluid flow domain. The flowrate for a system is likely to vary with time of day, seasons, and other factors. However, the geometry and configuration of a UV reactor system are typically fixed. Therefore, the flow field within a reactor system will also vary, although methods are available to accurately describe the flow field and its variations. The fluence rate field will be governed by the output power of the UV lamps in the system, the optical characteristics of the components of the system, and the geometry/configuration of the system. Analytical methods are available to measure each of the attributes of the system that influence the fluence rate field. As with the flow field, some of these attributes will vary with time. Additionally, as in the case of the flow field, it is possible to quantify the time-dependent behavior of each of these variables and, as such, it is also possible to define the time-dependent behavior of the fluence rate field of a reactor system.

In summary, system attributes that influence UV disinfection system performance can all be quantified independently. Methods are available to integrate results of these measurements to provide predictions of reactor behavior and these will be described in this chapter. When properly implemented, these methods allow accurate predictions of reactor behavior.

The validation protocols listed in Table 4.1 are all based on biodosimetry. Details of biodosimetric methods are presented in Chapter 3. These methods have the benefits of being familiar to most engineers and scientists who are involved in the design, validation, and regulation of UV systems. Another benefit of these methods is that they all are based on measured microbial inactivation. Biodosimetric methods also suffer from important drawbacks; the most important is that biodosimetry cannot be used to quantify the dose distribution delivered by a reactor system. Rather, system performance is characterized on the basis of a reduction equivalent dose (RED), which is a single-valued dose that is used to represent the inactivation behavior of the challenge organism itself. The RED value assigned to a reactor depends on the challenge organism that is used for the assay as well as the dose distribution delivered by the reactor for the conditions of the biodosimetric experiment. It is not possible to use the results of biodosimetry (i.e., RED) to provide quantitatively accurate predictions of the inactivation response of any other challenge or target organism, unless it has the same UV dose-response behavior as the challenge organism used in the biodosimetric assay. By extension, this requires the use of conservatism (safety factors) in translating the results of a biodosimetric validation test to the performance of the system relative to any other organism. As such, reactor validation by biodosimetry introduces conservatism that could be avoided by application of methods that account for the dose distribution.

3.0 CONTEMPORARY METHODS FOR ULTRAVIOLET REACTOR VALIDATION

In the mid-1990s, efforts were initiated to characterize and quantify the fundamental physical and chemical processes that govern the behavior of photochemical reactor systems; the concept of dose distribution was introduced by Blatchley and Hunt (1994). The first efforts to quantify the dose distribution in a reactor for a fixed operating condition were based on the use of laser Doppler velocimetry (LDV) to characterize the turbulent flow field (Chiu et al., 1999a). Process simulation was developed by interrogation of the measured flow field using a random-walk model to yield simulated trajectories of individual particles. In turn, a simulated fluence rate field was then mapped onto the simulated particle trajectories to yield estimates of trajectory-specific doses (Figure 4.2). This experiment-based method was successful at illustrating the concept of the dose distribution, and led to improvements in reactor design (Chiu et al., 1999b); however,

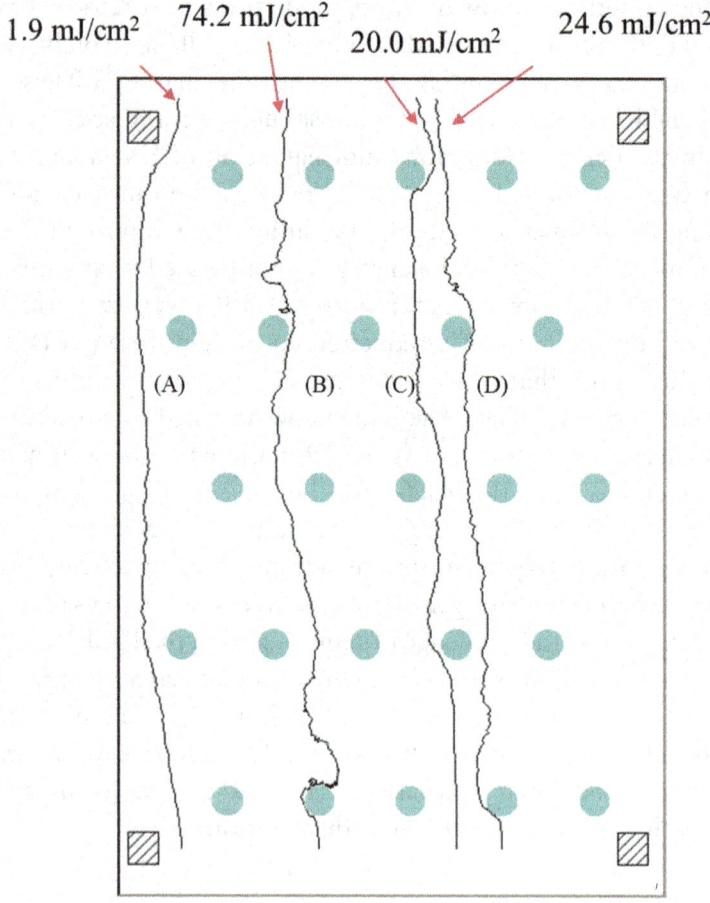

FIGURE 4.2 Simulated particle trajectories and corresponding trajectory-specific UV doses, as developed by mapping a numerical simulation of the fluence rate field onto simulated particle trajectories. In this case, particle trajectories were simulated by interrogation of measurements of the flow field by LDV (adapted from Chiu et al. [1999a]).

characterization of the flow field by LDV represents a time-consuming process that is difficult to justify for general applications. As a result, most efforts to estimate the dose distribution to date have involved the use of numerical models.

Computational fluid dynamics (CFD) represent a family of numerical methods that are used to simulate (turbulent) fluid flows in a wide range of engineering systems. The CFD-based simulations now form the basis for design of many engineering systems, including automobiles, aircraft, and reactor systems used in water treatment.

The CFD-based predictions of fluid flow in any system start from a digital, discretized representation (drawing) of the fluid flow domain (Figure 4.3). The domain defines the physical boundaries of the system and the region for which fluid flow will be simulated. Boundary conditions must be applied to the parts of the discretized domain that represent the boundaries of the region within which fluid motion is to be simulated, which typically include the entrance or

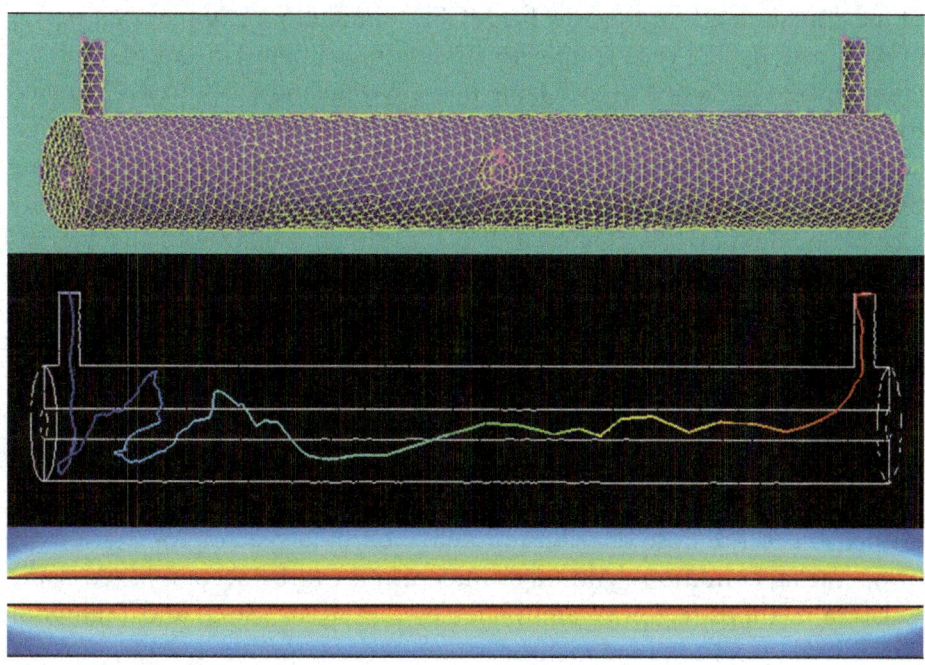

FIGURE 4.3 Illustration of components of a CFD-I simulation for a single-lamp UV disinfection system. The top panel illustrates discretized domain for CFD simulation; the middle panel illustrates a simulated trajectory of an individual particle through the domain (the simulated trajectory is developed by interrogation of a simulated [turbulent] flow field); the bottom panel illustrates a numerical simulation of the fluence rate (I) field for the domain. The F field is mapped onto each simulated trajectory and integrated, as described in eq 4.1, to yield an estimate of the trajectory-specific UV dose (figure adapted from Naunovic et al. [2005]).

exit of the system and interfaces between the moving fluid and (stationary) system boundaries, such as reactor walls or lamp jackets. For stationary surfaces within the computational domain, a no-slip boundary condition is typically applied. Entrance and exit nodes are typically assigned a boundary condition that defines a pressure or velocity condition. The simulation is conducted by developing approximate solutions to the equations of motion and continuity at the nodes of the discretized domain. Fluid mechanical behavior at intermediate locations is estimated by interpolation. Because fluid flow in virtually all contemporary UV reactors is in the turbulent range, it is also necessary to incorporate a turbulence closure model to account for the turbulence features of the flow field. A number of turbulence closure models and modeling approaches have been used to conduct these process simulations (e.g., Crapulli et al. [2010], Liu et al. [2007], Lyn and Blatchley [2005], and Sozzi and Taghipour [2006]).

The computational domain used for the fluence rate field model will represent the irradiated zone of the reactor system. This may include all or part of the domain used for the flow field (CFD) simulation. Lamp output power must be assigned for each lamp within the system. In most instances, identical lamp power is assigned to each lamp in the system and the power is assumed to be released uniformly along the length of the lamp axis. The optical characteristics of the system must also be assigned. A comprehensive fluence rate field simulation will include the transmittance characteristics of the quartz jackets and fluid in the system as well as the refractive indexes of all materials in the system. By following this approach, it is then possible to account for the factors that are known to influence the fluence rate field in a UV reactor system, that is, absorbance (Beer's law), dissipation, reflection, and refraction (Naunovic et al., 2008). Many numerical models and approaches have been applied to conduct fluence rate (I) field simulations and, to varying degrees, they are able to account for these physical phenomena.

The simulated flow field and fluence rate field are integrated to allow estimation of the dose distribution delivered by a reactor for a given set of operating conditions. Although integration can be accomplished using Eulerian methods (e.g., Lyn and Blatchley [2005] and Sozzi and Taghipour [2006]), the most common method is based on a Lagrangian frame of reference. In a Lagrangian simulation, the flow field is interrogated to allow for simulation of trajectories taken by individual particles through the irradiated zone of the system. The fluence rate field is then mapped on to each particle trajectory. Integration of the fluence rate history corresponding to each trajectory, as described by eq 4.1, allows for estimation of the dose corresponding to each trajectory. By repeating this process for a large number of particles, it is possible to develop a numerical approximation of the UV dose distribution delivered by a reactor for a given operating condition.

Subsequent integration of the dose distribution with the UV dose-response behavior of the target organism, as described by the following equation, allows for estimation of the inactivation response of the target organism to the dose distribution delivered by the reactor:

$$\left(\frac{N}{N_0}\right)_{reactor} = \int_0^\infty \left(\frac{N}{N_0}\right)_{batch} \cdot E(D) \cdot dD \qquad (4.2)$$

Where

$\left(\frac{N}{N_0}\right)_{reactor}$ = fraction of target organisms that retain viability downstream of the irradiated zone of the reactor,

$\left(\frac{N}{N_0}\right)_{batch}$ = UV dose-response behavior of target organism, and

$E(D)$ = UV dose distribution function.

Equation 4.2 is analogous to the segregated-flow model that is often used to describe the behavior of chemical and biochemical reactor systems (Levenspiel, 1972). The segregated-flow model is based on the assumption that reacting elements do not exchange material (i.e., they remain segregated) while traversing the active portion of a reactor system. When considering individual microorganisms within the irradiated zone of a UV reactor, it is evident that the assumptions of the segregated-flow model hold rigorously for UV disinfection systems. Equation 4.2 has also been demonstrated to be valid based on probabilistic arguments (Chiu et al., 1999a).

The dose distribution function is analogous to the residence time distribution function that is often used to describe the dynamic behavior of chemical or biochemical reactor systems (Levenspiel, 1972). As described in eq 4.2, the dose distribution takes the form of a density function. This requires the following characteristics of the dose distribution function:

$$\int_{-\infty}^{\infty} E(D) \cdot dD = \int_0^\infty E(D) \cdot dD \equiv 1 \qquad (4.3)$$

$$E(D) \cdot dD = \text{fraction of particles that receive dose} \qquad (4.4)$$
$$\text{between } D \text{ and } D + dD$$

Equations 4.3 and 4.4 define basic requirements of any function that is used to describe the dose distribution. Any method used for estimating the dose distribution must yield a function that satisfies these characteristics.

Lagrangian actinometry is the physical analog of the Lagrangian CFD-I modeling approach described previously. Lagrangian actinometry represents a method whereby the UV dose delivered to an individual particle can be measured. By applying the method to a population of particles, it then becomes possible to measure the UV dose distribution delivered by a reactor system.

All successful implementations of Lagrangian actinometry to date have involved the use of a cytidine derivative (*E*)-5-[2-(Methoxycarbonyl)ethenyl] cytidine (hereafter referred to as "*S*") as the actinometer. Bergstrom et al. (1982) reported that *S* undergoes a permanent photochemical change to yield 3-β-D-ribofuranosyl-2,7-dioxopyrido[2,3-*d*]pyrimidine (*P*) as a stable product via an unstable intermediate (see Figure 4.4). *P* is brightly fluorescent, whereas the parent compound (*S*) is nonfluorescent. Therefore, *S* and *P* can be distinguished by optical methods. In addition, the fluorescence response of *P* is quantitative. Moreover, the reaction illustrated by Figure 4.4 is photochemically efficient (i.e., it has a high quantum yield) for germicidally active UV radiation (Shen et al., 2005). Therefore, by conjugating large numbers of *S* molecules to the surface of microspheres, it becomes possible to quantify a range of UV exposures by individual microspheres within a population.

Microspheres selected for Lagrangian actinometry applications have size (diameter of 5 to 15 mm) and specific gravity ($\rho = 1.05$ g/cm^3) characteristics that closely mimic those of individual microorganisms. As such, the trajectories taken by the microspheres are representative of those taken by microorganisms as they traverse a UV reactor system. Therefore, the UV exposure histories and doses delivered to a population of these microspheres are representative of those delivered to a population of microbes as they traverse the same system under the same operating condition.

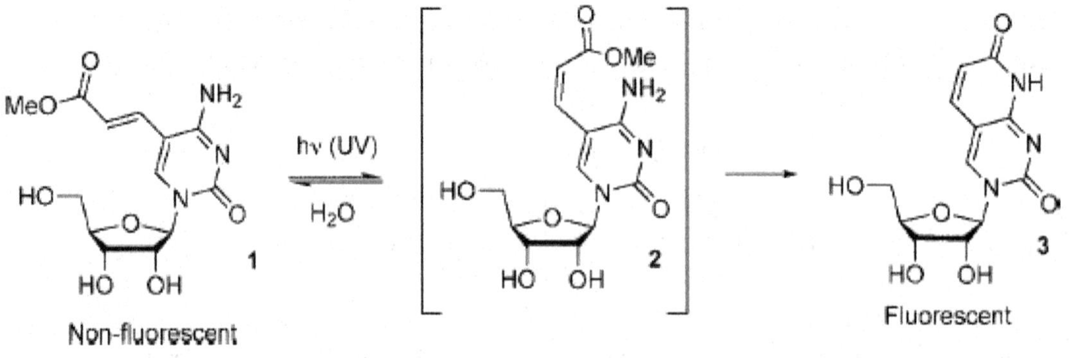

FIGURE 4.4 Photochemical transformation of *S* to yield *P*, as applied in Lagrangian actinometry. When subjected to exposure to germicidal UV radiation, the nonfluorescent compound *S* undergoes a reaction to yield an unstable compound (intermediate) that spontaneously decomposes to yield *P*, which is stable and brightly fluorescent.

In Lagrangian actinometry, a population of dyed microspheres is imposed on a UV reactor system that is being operated at a steady-state condition. The microspheres are collected downstream of the irradiated zone, then separated from water by particle–fluid separation processes, such as centrifugation and membrane filtration. The microspheres are then aspirated into a flow cytometer to allow measurement of the fluorescence intensity distribution for a large population of microspheres that has passed through the reactor.

In conjunction with this experiment, a subpopulation of the same microspheres is subjected to exposure to a range of UV doses under a collimated beam. The procedures used for microsphere exposure under the collimated beam are essentially identical to those used in biodosimetry. The microspheres from the collimated beam experiment are separated from water and analyzed using the same flow cytometry method to quantify the fluorescence intensity distributions that result from each dose within the collimated beam dose series.

The central hypothesis of the Lagrangian actinometry method is that the fluorescence intensity distribution of the reactor sample is the result of the convolution of the dose distribution and the fluorescence intensity distributions of the individual doses; this hypothesis has been confirmed (Blatchley et al., 2006). Thus, deconvolution methods can be used to estimate dose distribution delivered by a reactor. Deconvolution involves identification of the dose distribution that best describes the fluorescence intensity distribution observed in the reactor sample based on fluorescence intensity distributions measured in collimated beam samples. Deconvolution can be accomplished using algorithms in commercially available numerical modeling software packages such as MATLAB. Refinements of these methods have been developed that can yield improved accuracy for some circumstances (Cox et al., 2009).

Lagrangian actinometry has been shown to be effective for accurate measurement of the dose distribution delivered by continuous-flow UV reactors ranging from small point-of-use systems (Blatchley et al., 2006) to the largest UV reactors in the world (Blatchley et al., 2008). Lagrangian actinometry has also been shown to be effective for systems based on conventional low-pressure mercury lamps and systems based on medium-pressure mercury lamps (Shen et al., 2009). A protocol for application of Lagrangian actinometry has been developed (Scheible et al., Forthcoming).

Both CFD-I modeling and Lagrangian actinometry allow for estimation of the UV dose distribution delivered by a reactor. This attribute represents an important distinction from conventional biodosimetry, and has important implications with respect to reactor validation. Specifically, these methods have the potential to greatly reduce the uncertainty associated with predictions of reactor performance. An example of this is illustrated in Table 4.2, where the benefit of using the Lagrangian actinometry validation approach

TABLE 4.2 Estimated annual energy costs associated with reactor validation by biodosimetry (using coliphage T1 as the challenge organism) and those associated with validation by Lagrangian actinometry.

Average flowrate →		1600 m³/h (10 mgd)		16 000 m³/h (100 mgd)		160 000 m³/h (1000 mgd)	
UV transmittance (%)	P/Q (kW/mgd)	P (kW)	Annual cost ($/yr)	P (kW)	Annual cost ($/yr)	P (kW)	Annual cost ($/yr)
		Validation based on biodosimetry (challenge organism = Coliphage T1)					
95	0.329	3.3	2900	32.9	28,800	329	288,000
85	0.737	7.4	6500	73.7	64,600	737	646,000
75	1.25	12.5	10,900	125	109,000	1250	1,090,000
		Validation based on Lagrangian actinometry					
95	0.223	2.2	2000	22.3	19,500	223	195,000
85	0.500	5.0	4400	50.0	43,800	500	438,000
75	0.847	8.5	7400	84.7	74,200	847	742,000

Estimates of electrical power costs are based on an assumed unit cost of $0.10 per kW×hr. For each value of UV transmittance, a requirement of UV power per unit flowrate is calculated based on standardized protocols for biodosimetry and Lagrangian actinometry. Note that cost estimates in this table do not include additional savings in capital investment, components, operation and maintenance expenses, and labor. Electrical power cost estimates based on Lagrangian actinometry were developed based on application of dyed microspheres.

to reduce the uncertainty of reactor performance is reflected in terms of energy savings to different sizes of UV facilities.

4.0 EMERGING STRATEGY FOR REACTOR VALIDATION: STOCHASTIC APPROACH

Factors that are known to influence the behavior of UV disinfection systems are illustrated in Figure 4.1. All of these factors can be measured independently using well-defined analytical methods. In general, the nature of the dependencies among these factors is known. As such, it may be hypothesized that variability in overall performance is attributable to variability of the parameters that are known to affect process performance. Furthermore, it may be hypothesized that characterization of variability among the input parameters will allow for estimation of variability in overall process performance.

Figure 4.5 illustrates variability that is inherent in the UV dose-response behavior of *Escherichia coli* as a bacterial target organism. Also included in

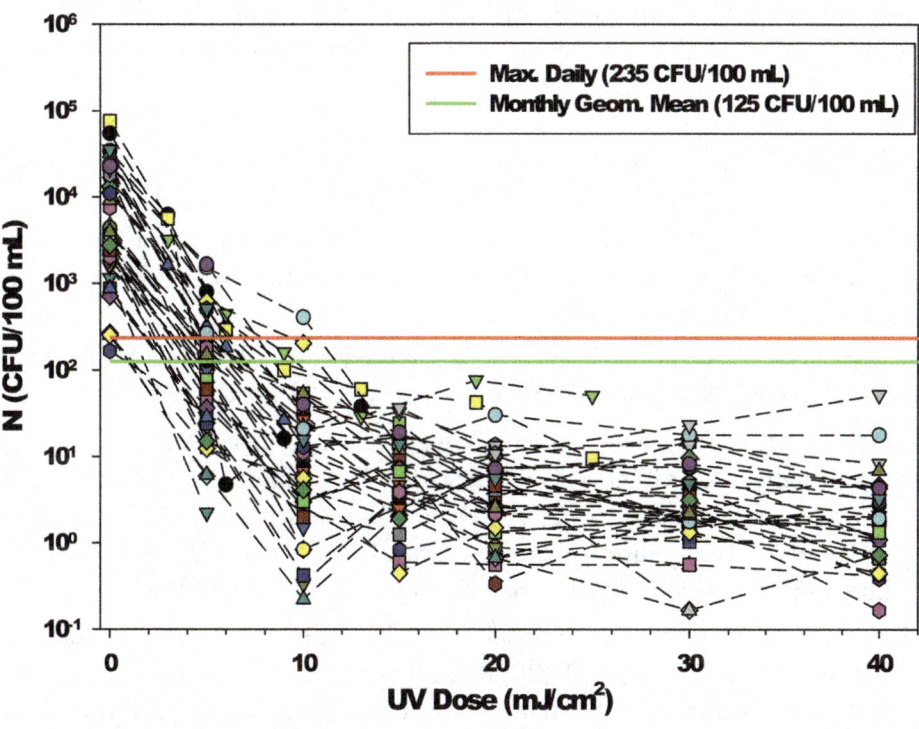

FIGURE 4.5 Measured variability on UV dose-response behavior of *E. coli* in a municipal wastewater effluent. Horizontal lines indicate typical National Pollutant Discharge Elimination System permit limits for viable *E. coli* in wastewater effluents (from Ortiz et al. [2013]).

this figure are common discharge permit limits that are applied for conventional disinfection applications (i.e., discharge of treated effluent to a receiving stream). These data indicate that collimated UV doses of 15 mJ/cm² are sufficient to achieve reliable compliance with the discharge permit limitation. These data also indicate that the value of N_0 (i.e., concentration of viable *E. coli* in undisinfected wastewater effluent) varies by roughly 3 orders of magnitude over the period of monitoring for this experiment and that, at the dose of 15 mJ/cm², variations of roughly 2 orders of magnitude are evident in the value of N (i.e., concentration of viable *E. coli* remaining after UV exposure).

The stochastic approach to simulation of the performance of UV disinfection systems, including variability, will involve detailed characterization of variability among each of the factors that is known to affect process performance. In turn, a CFD-I model will be used to simulate process performance in a UV disinfection system by allowing input process variables to vary over the ranges that are measured or simulated. By repeating these calculations in a stochastic sense, it is hypothesized that variability in process performance can be captured. If this is demonstrated to be correct, then the stochastic model could be used to inform design and operation of a UV disinfection system to minimize capital and operating expenses, while at the same time accomplishing reliable compliance with discharge permit limitations.

5.0 REFERENCES

Bergstrom, D. E.; Inoue, H.; Reddy, P. A. (1982) Pyrido[2,3-D]Pyrimidine Nucleosides: Synthesis via Cyclization of C-5-Substituted Cytidines, *J. Org. Chem.*, **47** (11), 2174–2178.

Blatchley, III, E. R.; Hunt, B. A. (1994) Bioassay for Full-Scale UV Disinfection Systems. *Water Sci. Technol.*, **30** (4), 115–123.

Blatchley, III, E. R.; Shen, C.; Naunovic, Z.; Lin, L.; Lyn, D. A.; Robinson, J. P.; Ragheb, K.; Grégori, G.; Bergstrom, D. E.; Fang, S.; Guan, Y.; Jennings, K.; Gunaratna, N. (2006) Dyed Microspheres for Quantification of UV Dose Distributions: Photochemical Reactor Characterization by Lagrangian Actinometry. *J. Environ. Eng.*, **132** (11), 1390–1403.

Blatchley, III, E. R.; Shen, C.; Scheible, O. K.; Robinson, J. P.; Ragheb, K.; Bergstrom, D. E.; Rokjer, D. (2008) Validation of Large-Scale, Monochromatic UV Disinfection Systems Using Dyed Microspheres. *Water Res.*, **42** (3), 677–688.

Bolton, J. R.; Linden, K. G. (2003) Standardization of Methods for Fluence UV Dose Determination in Bench-Scale UV Experiments. *J. Environ. Eng.*, **129** (3), 209–215.

Chiu, K.; Lyn, D. A.; Savoye, P.; Blatchley, III, E. R. (1999a) An Integrated UV Disinfection Model Based on Particle Tracking. *J. Environ. Eng.*, **125** (1), 7–16.

Chiu, K.; Lyn, D. A.; Savoye, P.; Blatchley, III, E. R. (1999b) Effect of System Modifications on Disinfection Performance: Pilot Scale Measurements and Model Predictions. *J. Environ. Eng*, **125** (5), 459–469.

Cox, E. M.; Xia, J.; Craig, B.; Shen, C.; Scheible, O. K.; DiToro, D.; Blatchley, III, E. R. (2009) Numerical Methods in Lagrangian Actinometry: Solving for a UV Reactor's Dose Distribution. *Proceedings of Disinfection 2009*; Atlanta, Georgia, Feb 28–March 3; Water Environment Federation: Alexandria, Virginia.

Crapulli, F.; Santoro, D.; Haas, C.N.; Notarnicola, M.; Liberti, L. (2010) Modeling Virus Transport and Inactivation in a Fluoropolymer Tube UV Reactor Using Computational Fluid Dynamics. *Chem. Engr. J.*, **161**, 9–18.

DVGW (2006) *UV Devices for the Disinfection of the Water Supply*; German Standard W 294-1, 294-2, 294-3.

Great Lakes Upper Mississippi River Board (GLUMRB) (2004) *Recommended Standards for Wastewater Facilities*. http://10statesstandards.com/wastewaterstandards.html (accessed Feb 2015).

International Ultraviolet Association (2010) *IUVA Manufacturers Council Position on a Uniform Protocol for Wastewater UV Validation Applications*; International Ultraviolet Association: Washington, D.C.

International Ultraviolet Association (2012) *Low Dose UV Systems for Secondary Effluent Applications*; IUVA Manufacturers Council.

Levenspiel, O. (1972) *Chemical Reaction Engineering*; Wiley & Sons: New York.

Liu, G.; Slawson, R. M.; Huck, P. M. (2007) Impact of Flocculated Particles on Low Pressure UV Inactivation of *E-coli* in Drinking Water. *J. Water Supply. Res. T.*, **56**, 153–162.

Lyn, D. A.; Blatchley, III, E. R. (2005) Numerical Computational Fluid Dynamics-Based Models of Ultraviolet Disinfection Channels. *J. Environ. Eng.*, **131** (6), 838–849.

National Water Research Institute; Water Research Foundation (2012) *Ultraviolet Disinfection Guidelines for Drinking Water and Water Reuse*, 3rd ed.; National Water Research Institute: Fountain Valley, California.

Naunovic, Z.; Pennell, K.; Blatchley, III, E. R. (2008) The Development and Performance of an Irradiance Field Model for a Cylindrical Excimer Lamp. *Environ. Sci. Technol.*, **42** (5), 1605–1614.

Naunovic, Z.; Shen, C.; Lyn, D. A.; Blatchley, III, E. R. (2005) Modeling and Design of an Ultraviolet Water Disinfection System. *SAE J. Aerospace*, 554–563.

ÖNORM (2001) *Plants for Disinfection of Water Using Ultraviolet Radiation Requirements and Testing: Low Pressure Mercury Lamp Plants*; Austrian Standard 5873-1.

Ortiz, A. P.; Chiu, D.; Grady, C.; Blatchley, III, E. R. (2013) Variability in the Performance of Municipal Wastewater UV Disinfection Systems. *Proceedings of the IOA and IUVA World Congress*, Las Vegas, Nevada, Sept 24.

Scheible, O. K.; Shen, C.; Blatchley, E. R. (Forthcoming) Protocol for Validation of UV Disinfection Reactors Using Lagrangian Actinometry; Water Research Foundation: Denver, Colorado.

Shen, C.; Fang, S.; Bergstrom, D. E.; Blatchley, III, E. R. (2005) (*E*)-5-[2-(methoxycarbonyl)ethenyl]Cytidine as a Chemical Actinometer for Germicidal UV Radiation. *Environ. Sci. Technol.*, **39**, 10, 3826–3832.

Shen, C.; Scheible, O. K.; Chan, P.; Mofidi, A.; Yun, T. I.; Lee, C. C.; Blatchley, III, E. R. (2009) Validation of Medium-Pressure UV Disinfection Reactors by Lagrangian Actinometry Using Dyed Microspheres. *Water Res.*, **43**, 1370–1380.

Sozzi, D. A.; Taghipour, F. (2006) UV Reactor Performance Modeling by Eulerian and Lagrangian Methods. *Environ. Sci. Technol.*, **40**, 1609–1615.

U.S. Environmental Protection Agency (2006) *Ultraviolet Disinfection Guidance Manual for the Final Long Term 2 Enhanced Surface Water Treatment Rule*; EPA-815/R-06-007; U.S. Environmental Protection Agency: Washington, D.C.

Process Design and System Sizing

Andrew Salveson, P.E.; Keith Bourgeous, Ph.D., P.E.; Nicola Fontaine; Norayo Noibi; and Bill Sotirakos

1.0 DISINFECTION PERMIT REQUIREMENTS	96	
2.0 WASTEWATER QUALITY EFFECTS ON ULTRAVIOLET DISINFECTION	97	
2.1 Effluent Water Quality	97	
2.1.1 Dissolved Constituents and Their Effect on Ultraviolet Absorbance and Transmittance	98	
2.1.1.1 Dissolved Organic Matter	98	
2.1.1.2 Inorganic Compounds	99	
2.1.2 Particles	100	
2.2 Water Quality Characterization Tools	101	
2.2.1 Turbidity	103	
2.2.2 Total Suspended Solids	103	
2.2.3 Ultraviolet Transmittance	106	
2.2.3.1 Seasonal and Diurnal Ultraviolet Transmittance Variability	106	
2.2.3.2 Secondary Process Effects on Ultraviolet Transmittance	106	
2.2.3.3 Chemical Effects on Ultraviolet Transmittance	107	
2.2.3.4 Industrial Effects on Ultraviolet Transmittance	108	
2.2.3.5 Sidestream Flow Effects on Ultraviolet Transmittance	109	
2.3 Upstream Processes to Improve Water Quality	110	
2.4 Fouling of Lamp Sleeves, Lamp Racks, and Channels	113	
3.0 REACTOR SELECTION CRITERIA	114	
3.1 System Configurations	114	
3.1.1 Closed Vessel	114	
3.1.2 Open Channel	115	
3.1.3 Lamp Orientation	115	
3.1.4 Non-Submerged Ultraviolet Lamp Systems	115	
3.1.5 Lamp Spacing	115	

3.2	Establishing Design Criteria	116	3.3.2.1 Dose Pacing Based on Flow, Ultraviolet Transmittance, and Power Setting	120
	3.2.1 Flow	116		
	3.2.2 Headloss and Water Level	117		
	3.2.3 Influent and Effluent Water Quality	118	3.3.2.2 Dose as a Function of Flow, Ultraviolet Transmittance, and Ultraviolet Intensity	120
3.3	Design Dose and Dose Control Strategies	118		
	3.3.1 System Monitoring	119	4.0 ULTRAVIOLET SYSTEM SIZING EXAMPLE	120
	3.3.2 Dose Delivery Strategies	119	5.0 REFERENCES	124

This chapter provides a review of UV reactor process design, starting with permit requirements, addressing water quality issues that affect UV performance, and highlights UV reactor design approaches. Additionally, this discussion is intended for disinfection of secondary and tertiary effluents and does not represent a manual for water reuse treatment, some effluent discharges have similar permit requirements and, as such, the information provided herein may be of use for these low-level water reuse applications.

1.0 DISINFECTION PERMIT REQUIREMENTS

The first step in disinfection design is to determine the applicable disinfection permit requirements, both existing and projected. As described in Chapter 1, there are a range of disinfection standards, which depend on discharge location and degree of public contact. As noted in Chapter 1, permit limits are generally based on water quality standards or criteria that are protective of human health. Implementation of these standards are based on indicator parameters that have been shown to be linked to specific levels of human health risk. Thus, wastewater permits are based on indicator organisms such as fecal coliform, *Escherichia coli*, or enterococcus for the practical purposes of permit implementation and monitoring. Wastewater discharge permits require disinfection of bacteria; UV disinfection that is designed to meet these limits is also effective at providing disinfection of other pathogenic organisms, including viruses and protozoa, as described in Chapter 2. Thus, unlike drinking water, wastewater UV disinfection systems are typically designed around meeting bacterial indicator standards. The following are examples of wastewater disinfection requirements in the United States:

- A utility in the state of Washington that discharges to surface water has an effluent disinfection limit based on fecal coliform (monthly geometric mean of 200 colony-forming units [cfu]/100 mL and a weekly geometric mean of 400 cfu/100 mL); the monitoring frequency for fecal coliform is three times per week;
- A utility in the state of Colorado that discharges to a river has an effluent disinfection limit based on *E. coli* (monthly geometric mean of 126 cfu/100 mL); the monitoring frequency of *E. coli* in the utility effluent is once per month; and
- A utility in Texas that discharges into the Gulf of Mexico has an effluent disinfection limit based on enterococci (monthly geometric mean of 35 cfu/100 mL and a daily maximum of 89 cfu/100 mL); monitoring frequency is five times per week.

It is important to note that compliance for bacteria is typically based on geometric means and, from a permit compliance standpoint, a geometric mean allows for a limited number of higher effluent values without affecting permit compliance; however, many permits also have a daily maximum requirement that sets an upper limit on these values.

2.0 WASTEWATER QUALITY EFFECTS ON ULTRAVIOLET DISINFECTION

One of the most significant considerations in design of a UV disinfection system is understanding the effect of water quality on both the efficacy and efficiency of UV disinfection. The following sections include a summary of research on our understanding of how various parameters affect UV disinfection; the summaries are supported by site-specific examples of these factors.

2.1 Effluent Water Quality

Treated wastewater effluent includes both dissolved constituents and particulates; particles may be inorganic or organic. Inorganic particles include iron and alumina oxides and dissolved inorganics include other ions such as calcium, magnesium, or carbonates that contribute to water hardness or alkalinity. Biological–organic living particles include viruses, bacteria, and algae, while nonliving biological particles include cellular debris (Wilkinson et al., 1997), or dead microbial cells. Particles may range in size from a few nanometers on the border between dissolved and colloids, submicron, micron range, and up to millimeter dimensions, as with sand particles. A distinction is drawn between colloidal and suspended particles, with an arbitrary boundary of 1 or 0.45 μm. Colloidal particles vary between 0.001 and 1 μm, and dissolved constituents are typically smaller than 0.001 μm; however, the distinction may depend on the quantification method of colloids and dissolved constituents.

2.1.1 Dissolved Constituents and Their Effect on Ultraviolet Absorbance and Transmittance

Ultraviolet photolysis is effective when the target molecule absorption spectrum overlaps the emission spectrum of the UV lamp and when the quantum yield is reasonably large. Inorganic and organic matter present in water can affect UV disinfection efficiencies by absorbing UV light, thus limiting the light available for inactivation of bacteria. Dissolved organic and inorganic substances can be evaluated by measurement of total dissolved solids. Ultraviolet transmittance (UVT) percentage is a general measure of the fraction of light (at 254 nm) transmitted through a 1-cm water sample and if often used to predict UV efficiency by allowing estimation of the photons available for reaction with target microorganisms. Higher UVT means lower water absorbance and higher predicted UV efficiency.

2.1.1.1 Dissolved Organic Matter

Dissolved organic matter (DOM) present in wastewater effluent is a complex matrix of various organic molecules typically comprised of biogenic, polyelectrolytic, organic molecules and polymers, with mass concentration ranging from 0.5 to 100 mg/L of organic carbon. Dissolved organic matter has been categorized as humic substances and nonhumic substances (Frimmel, 1998). Humic substances can absorb light in the UV and visible range mostly because of unsaturated structures (Frimmel, 1998). Dissolved organic matter affects UV process efficiency through three main mechanisms:

- Dissolved organic matter can absorb light in both the UV and the visible ranges, hindering treatment;
- Photosensitization reactions from DOM by absorption of UV light can result in formation of triplet excited states, "3DOM*", that react further with O_2 to generate reactive oxygen species. Lester et al. (2013) validated and developed methods for determining the quantum yields of photo-oxidants ($\cdot$OH, $^1O_2$, 3DOM*, and H_2O_2) generated by DOM under 254 nm UV C radiation; and,
- Dissolved organic matter can act as a radical inhibitor or a radical promoter, depending on its functional groups and molecular size (Bianchini et al., 2002; Pereira et al., 2007; Metz et al., 2011).

Another phenomenon that can affect disinfection is sorption of DOM to surfaces (e.g., particles and bacteria). Organic macromolecules may adsorb to inorganic colloids, forming a thin layer at the colloid surface and modifying the interaction energy barrier (Buffle and Leppard, 1995). Suspended particles can absorb DOM in wastewaters and change their surface properties (Dai and Hozalski, 2002). Sorption of organic matter on colloids or larger particles may affect UV disinfection and DOM may be used as an additional nutrient carbon source for the surviving bacteria following UV disinfection and enhance bacterial regrowth (Camper et al., 2001; Metz et al., 2011), which was previously described in Chapter 2.

2.1.1.2 Inorganic Compounds

Inorganic compounds such as iron can occur naturally, or metal salts may be added during wastewater treatment processes. For example, nitrate and nitrite, which may be present in nitrified or partially nitrified effluents, are strong UV absorbers at wavelengths below 240 nm (Guenther et al., 2001) and may strongly affect efficiency of polychromatic medium-pressure UV systems (Sharpless et al., 2003). Both nitrate and nitrite can form hydroxyl radicals (·OH) when exposed to UV light; thus, photolysis of these compounds may be advantageous (Keen et al., 2012). Nitrite is typically not present in significant concentrations in secondary wastewater effluents unless there are operational issues in the biological treatment process, but it may also form as a byproduct of nitrate photolysis; nitrate is often present at greater than 5 mg/L nitrate-N.

When iron is used in the treatment process, it can precipitate from solution to form UV-absorbing ferric floc particles, which can affect disinfection by forming coagulated particles. However, dissolved iron remaining in solution can also have an effect on UV absorbance, which can also affect disinfection efficiency. Nourmoradi et al. (2012) showed that increasing an iron concentration from 0.1 to 0.5 mg/L led to a decrease (0.3 to 0.5 log) of the inactivation efficiency of UV radiation of *Aspergillus* spp. in water. This is consistent with the fact that iron decreases UVT; impact threshold concentrations, which are concentrations that result in UVT decreases from 91 to 90%, have been reported at 0.057 mg/L for ferric iron ($Fe2^+$) and 9.6 mg/L for ferrous iron ($Fe3^+$) (Bolton et al., 2001). If the UV system is being designed for an existing facility, data on iron concentrations could be collected to determine if they exceed threshold values. If the facility is part of a new water resource recovery facility, consideration should be given to selection of metal salts for coagulation ahead of UV disinfection systems.

Iron and other dissolved inorganics may also affect UV disinfection if these compounds lead to fouling on lamp sleeves, which reduces the transmittance of UV light through the sleeve into the water. Also, fouling on sensor windows can affect measured UV intensity and dose monitoring. Sleeve fouling can be accounted for with the fouling/aging factor in the design of the UV facility. While inorganic fouling is a complex process, it is known that it is related to hardness, iron, and other inorganic constituent concentrations. The solubility of inorganic constituents depends on whether they are in an oxidized or reduced state, which can be affected by both the oxidation–reduction potential and pH of the water (Wait and Blatchley, 2010). Predicting the fouling potential with regard to iron and other inorganic foulants can be challenging. However, Wait and Blatchley (2010) demonstrated that iron is the most dominating metal foulant in UV reactors, with elevated ability to absorb UV, even when calcium is the most predominant metal in groundwater. Manganese, aluminum, and zinc are generally present at low concentrations, and contribute a small amount to lamp sleeve fouling. Five- to 12-month pilot studies using UV reactors with low-pressure, low-pressure high-output (LPHO), and medium-pressure lamps have shown that standard cleaning protocols and wiper frequencies (1 to 12 cleaning cycles per hour) are generally sufficient to overcome the effect of sleeve fouling with water that had total and calcium hardness levels less than 140 mg/L and iron less than 0.1 mg/L (Mackey et al., 2001; Mackey et al., 2004).

2.1.2 Particles

The dose-response curve of dispersed free-swimming microorganisms is expected to follow first-order kinetics for UV inactivation. A lag or shoulder effect is obtained at low doses because of microbial self-aggregation (Bohrerova et al., 2006; Mamane-Gravetz and Linden, 2004; Severin et al., 1983) or cellular repair mechanisms (Jagger, 1967), and a reduced inactivation, termed the *tailing effect*, may be obtained at high doses because of particle effects (Loge et al., 2001; Qualls et al., 1983). Numerous studies have shown that microbial pathogens can associate with particles and this particle association can affect the efficacy of UV disinfection.

With respect to bacteria, research by Scheible (1987) has shown that coliform bacteria can associate with particles and be shielded from UV light. However, some UV light can penetrate wastewater flocs because of the porosity of the particles or light-accessible pathways that create channels for light penetration. Ultraviolet light is unable to penetrate even a few microns within a floc directly because of the high absorbance of wastewater particles sufficient to block UV light caused by absorption and scattering (Emerick et al., 2000; Loge et al., 1999; Qualls et al., 1983). A residual microbial concentration observed at high UV doses is indicative of a system where microbes are shielded from UV light. Inactivation of bacteria in wastewater effluent may not reach above 4- to 5-log inactivation because of particle–microbe interactions. These interactions result in transition to the tailing zone of the dose-response curve at doses that are typically between 20 to 40 mJ/cm^2 (Bohrerova and Linden, 2006; Gehr and Nicell, 1996; Gehr et al., 2003; Guo et al., 2011; Loge et al., 2001).

Interestingly, viruses in secondary effluent do not appear to associate strongly with particulate matter. Ho et al. (1998) investigated the inactivation of indigenous male-specific coliphages and showed that inactivation efficiency was not significantly reduced when mixed liquor suspended solids was spiked to simulate storm events with total suspended solids (TSS) levels between 10 and 60 mg/L. The residual phage seemed to have no correlation with the TSS level, and good inactivation was achieved even at high TSS. However, Templeton et al. (2005) reported that activated sludge particles can enmesh and protect MS2 coliphage and bacteriophage T4 from low-pressure UV disinfection; as such, additional research may be needed to more fully characterize the association of pathogenic viruses with wastewater particles. In any case, depending on the target value for treatment performance, wastewater particles can affect UV disinfection. And, there are a number of methods that can be used to characterize particles in wastewater effluents.

Thus, disinfection effectiveness is affected by the presence and characteristics of particles, and characterization of suspended solids has become important information in designing UV disinfection systems (Metcalf and Eddy, 2014). There are two main effects that particulate matter has on UV disinfection efficacy and efficiency. First, particulate matter in water interferes with transmission of UV light. Second, the effect of particles on UV disinfection also includes shielding (shading) and enmeshment (embedding). Figure 5.1a illustrates the effect of particle "shielding" on UV disinfection by absorption, scattering, refraction, and reflection. Figure 5.1b illustrates the effect of particles enmeshed or associated with microorganisms on UV disinfection, where microbial inactivation depends on accessible pathways to UV light; this is a phenomenon that has been studied mainly with UV disinfection of wastewater effluents.

2.2 Water Quality Characterization Tools

To develop criteria for the design of disinfection systems that address the issues associated with water quality characteristics discussed previously, it is important to develop information on both the dissolved and particulate components of the treated effluent. The characterization of effluent with respect to the effect of dissolved constituents can be determined through evaluation of UVT data. Ultraviolet transmittance as a parameter was described

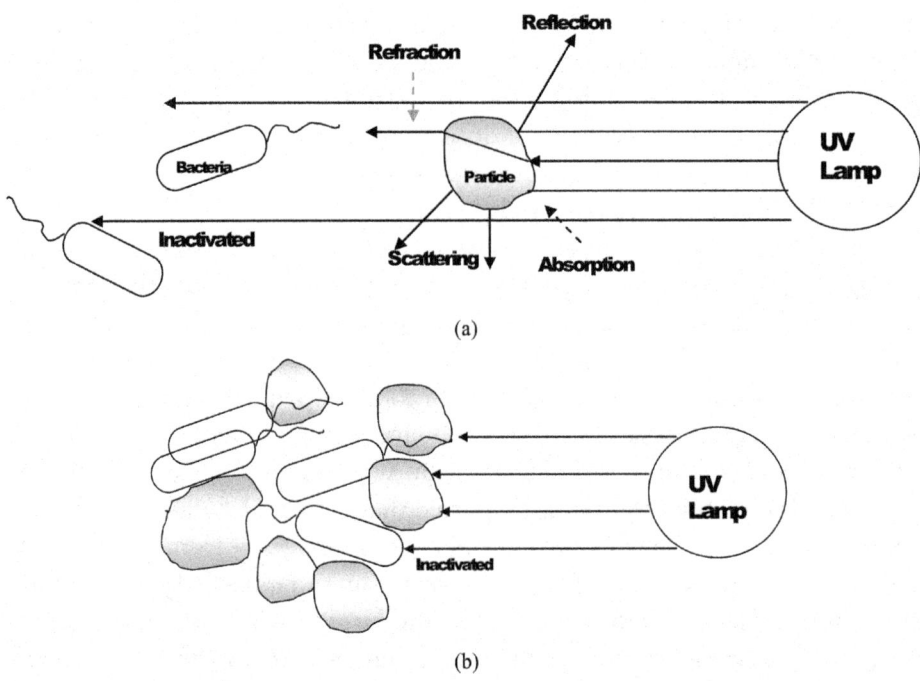

FIGURE 5.1 (a) The effect of particle "shielding" on UV disinfection and (b) the effect of particles enmeshed/associated with microorganisms on UV disinfection.

in Chapter 2; it reflects the efficiency of the ability of the water to pass UV light. While simply measuring UVT provides an indication of the UVT that will be treated by a given system, it may also be useful to laboratory-filter samples and repeat the UVT measurements to determine whether the results could be improved by additional removal of solids. Conducting a laboratory filtration of a treated sample will allow the design engineer to determine whether a depressed UVT is the result of light scattering from particles or from absorption of UV light from dissolved constituents in the water.

Light scattering by particles in water can be measured using turbidity, and it is one of the most common tests used in drinking water and filtered effluent. Turbidity provides information on the colloidal particles (<1 μm) and thus can be a measure of colloidal removal. Total suspended solids represent the portion of total solids retained by a specified filter, measured after being dried. Both turbidity and TSS are "lumped" parameters that do not provide information on particle size, shape, or their concentration, and thus may not be the best measure of particle-removal efficiency. Nevertheless, TSS is a universally used wastewater effluent standard (along with biochemical oxygen demand [BOD]) to evaluate performance of conventional water resource recovery facilities (WRRFs) for regulatory control. There is no relationship between turbidity and TSS in untreated wastewater, however, a reasonable relationship may exist in filtered secondary effluent from activated sludge and it is facility-specific (Metcalf and Eddy, 2014).

There are other methods used to characterize particles such as microscopy methods including imaging techniques; particle counters (light interaction methods, electrical property methods, and automated image analysis); and separation methods (sedimentation, centrifugation, serial membrane filtration, fractionation, and classification). Particle size analyzers measure particle size distribution, that is, the fraction of particles within a certain particle diameter size range. Particles and flocs in wastewater have irregular shapes that exist simultaneously and differ in size (Shubert and Gunthert, 2001). This section will provide an overview of commonly used methods for characterizing particulates and UVT.

2.2.1 Turbidity

The light-scattering effect of disperse inorganic particles is mostly relevant in UV inactivation of microorganisms in water, while the effect of particle–microbe association in effluent is mostly relevant in wastewater (U.S. EPA, 2006). In drinking water, studies showed that the influence of turbidity up to values of 10 nephelometric turbidity units on UV inactivation was insignificant (Amoah et al., 2005; Mamane, 2008). In studies where coagulation was used, Liu et al. (2007) demonstrated that floc particles formed by coagulation and flocculation led to lower (more than 1-log) inactivation of *E. coli* at 10 to 40 mJ/cm^2 compared to non-particle-associated *E. coli* at similar turbidity values. Based on these findings, it is apparent that turbidity measurements do not provide information on the extent of association of microbes with particles. Unlike suspended solids concentrations, which can be correlated to some extent with aggregation (although site-specific), turbidity is not an effective method for estimating the effect of solids on UV disinfection performance in secondary wastewater applications.

2.2.2 Total Suspended Solids

Bacteria that exist in wastewater effluent may exist as free-living cells or be associated with particles via aggregation or floc. Aggregates or flocs are significantly different from their constituent particles and microorganisms in size, shape, porosity, density, and composition (Droppo, 2001). Activated sludge bioflocs are a polymeric network of microorganisms (mainly bacteria, but also organisms such as fungi, viruses, and protozoa); extra-cellular polymeric substances (EPS); inorganic particles (as calcium phosphate and iron oxides); divalent cations (Urbain et al., 1993); and bound and free water within the floc pores (Droppo, 2001). Microorganisms can naturally form aggregates among themselves (i.e., self-aggregation) and with other particles because of favorable environmental conditions such as improved food assimilation and protection from environmental stresses (Gerba and McLeod, 1976). Microorganisms can also aggregate with particles through artificially enhanced processes by chemical coagulation and flocculation. Loge et al. (1999) determined chemically flocculated particles (i.e., activated sludge with chemical phosphorus removal) to be denser then biological flocs produced by aeration tanks, thus shielding UV light into the aggregate to a greater extent. Addition of polymeric flocculants to the chemicals used to induce chemically flocculated flocs may form large and low-density flocs (Gregory, 1997) that can affect disinfection. Particles associated with coliforms in wastewater are above 10 μm in diameter; this depends on the type of upstream biological process (Emerick et al., 2000).

Interestingly, while it is used widely as a general rule for indicating whether an effluent is amenable to UV disinfection, TSS alone may not be an appropriate parameter for evaluating UV disinfection performance (Madge and Jensen, 2006; Nelson, 2000; Qualls et al., 1983). Ho et al. (1998) found that the residual total coliform after UV inactivation did not have a strong correlation with TSS. At high UV doses, a good inactivation was achieved regardless of the TSS concentration. Conversely, Scheible (1987) found that residual effluent coliforms are significantly influenced by TSS, while Beltran and Jimenez (2008) showed that UV inactivation of free fecal coliforms was not dependent on the TSS value; in addition, inactivation rates in the tailing region decreased with an increase in TSS values between 15 to 100 mg/L. Ho et al. (1998) reported that there is no correlation between TSS and the number of particles containing coliforms and suggested that the degree in which coliforms are embedded in particles determines the level of inactivation achieved. Nelson (2000) examined the disinfection of wastewater stabilization ponds and found that, despite high TSS, the pond effluent was easily disinfected with UV light because the majority of the coliforms were not associated with particles. Therefore, it is not simply the presence of solids that affects disinfection efficacy, rather, the level of association of bacteria within the particles. Thus, it is often useful to conduct additional characterization of particles that may include particle numbers and size distribution.

Figure 5.2 shows particle size distribution graphs of secondary effluents generated from different types of biological treatment processes, demonstrating that the type and mode of the upstream biological treatment process affects the size and number of particles present in a secondary effluent. Effluents generated from ponds, pure oxygen activated sludge, and trickling filters have higher particle concentrations than effluents generated from air activated sludge or oxidation-ditch plants. In addition to the presence of larger particles being a concern, the content of these particles

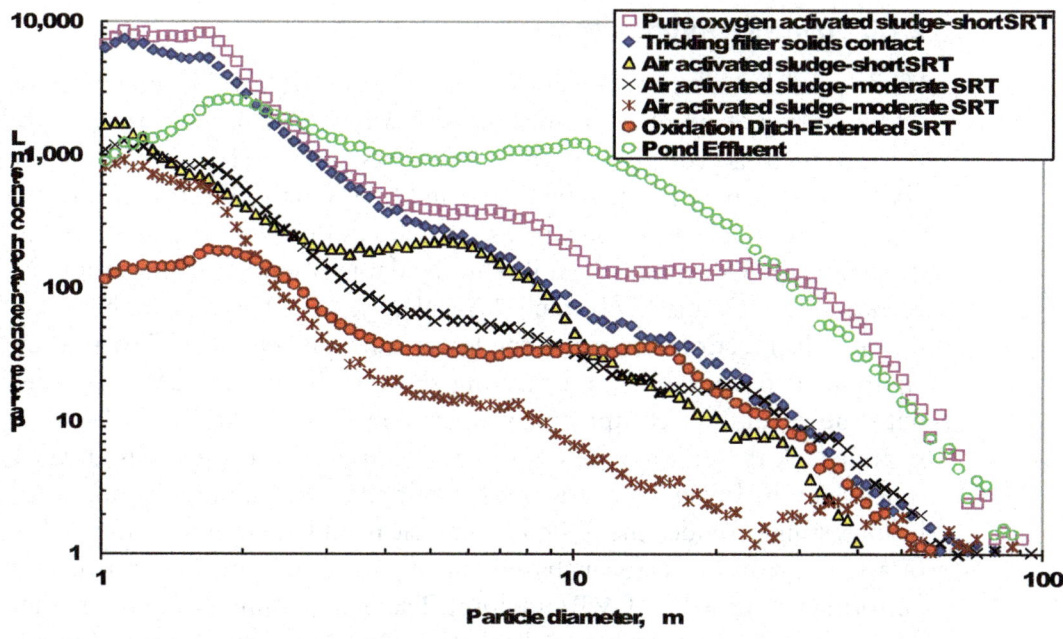

FIGURE 5.2 Particle size distribution of several biological treatment processes.

can have an effect. As a part of Water Environment Research Foundation Project 96-CTS-3 (Darby et al., 1999), it was also determined that the type of biological treatment process affects the content within the wastewater particles. As shown in Table 5.1, WRRFs that use pure oxygen activated sludge or trickling filter biological treatment processes tend to produce a higher percentage of coliform-associated particles.

TABLE 5.1 Effect of treatment process on particle-associated bacteria.*

Treatment process	Percentage of particles containing coliform bacteria
Activated sludge with pure O_2	3.1–20.2
Trickling filter	13–28
Activated sludge with air	1.1–11.5
Trickling filter/solids contact	3
Facultative pond	None detected

*From the effect of upstream treatment processes on UV disinfection performance by Darby et al. (1999)

2.2.3 Ultraviolet Transmittance

While particles affect disinfection efficacy, UVT affects the economics of UV operation. A number of substances in water influence UV absorbance directly such as iron, nitrate, and DOM, as previously described. Low UVT waters can be effectively disinfected by UV light, but this will result in a higher cost. Ultraviolet transmittance data should be analyzed to determine the value and variability in the data set, and the data should be used to set the design UVT value. Typical design UVT values are based on meeting disinfection criteria at the lower 10th percentile UVT with duty equipment in operation and meeting the remaining low UVT episodes with the redundant equipment in operation. The variability in UVT, or relative lack thereof, may allow the design engineer to use a design UVT other than the lower 10th percentile. Coupling UVT data with collimated beam testing provides the design team a clear understanding of the ability of a UV system to perform (based entirely on dose) and the cost of that performance (based on UVT and dose). There are a number of factors that are known to affect UVT, and their effects on design are described in the following sections.

2.2.3.1 Seasonal and Diurnal Ultraviolet Transmittance Variability

Ultraviolet transmittance of an effluent can vary by season; the reasons behind seasonal variations often are attributable to changes in secondary process operation (i.e., different levels of biological treatment because of seasonal permit variance), seasonal industrial discharge (such as agricultural discharges), and dilution during large storm flow/inflow events. Understanding variation in UVT seasonally is critical to achieving permit compliance. Figure 5.3 shows the variation of the secondary effluent UVT at the Laguna WRRF in Santa Rosa, Californa, between October 2013 and April 2014; UVT varies between 49 and 75%. The selected UV equipment must be capable of providing the target dose to the wastewater flow over the expected UVT range. The design team should consider variability in UVT during equipment selection and system design because UVT affects system capacity and cost.

2.2.3.2 Secondary Process Effects on Ultraviolet Transmittance

Although there are many factors upstream of the WRRF (i.e., potable water quality, commercial and industrial discharges, and collection system infiltration) that can affect the UVT of a secondary effluent, the type and mode of operation of the biological treatment process can also affect the effluent UVT. A full-scale comparison of secondary effluent UVT values

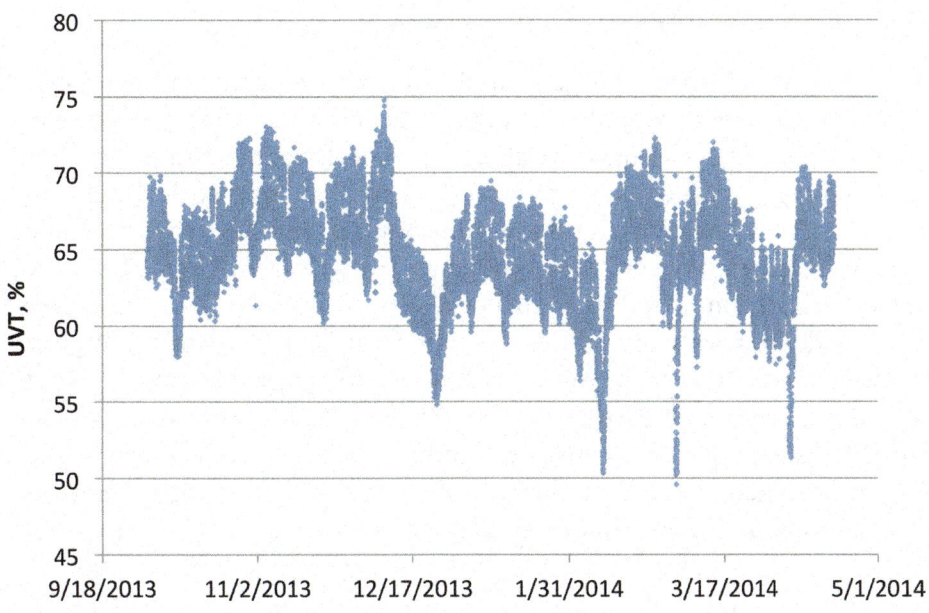

FIGURE 5.3 Variation in UVT at Santa Rosa, California.

generated by an activated sludge and pond system treating the same influent wastewater from the Napa Sanitation District (Napa, California) shows that the effluent UVTs are 74 and 39%, respectively. There are exceptions to this observation, but, generally, secondary effluents from attached growth processes (i.e., trickling filters) and some pure oxidation activated treatment facilities tend to produce an effluent with a lower UVT value than other treatment facilities.

2.2.3.3 Chemical Effects on Ultraviolet Transmittance

Ultraviolet transmittance decreases as a result of UV-absorbing substances (organic or inorganic) and particles that either absorb or scatter UV light. As part of upstream wastewater treatment, some UV-absorbing chemicals are used. The effects of these chemicals may be observed further downstream in the treatment process; for example, ferric chloride, typically used for coagulation and flocculation for phosphorus removal or for odor control, can have negative effects on UVT and can deposit on quartz sleeves (i.e., foul). Declines in UVT increase the energy requirements to maintain a UV dose. Other chemicals used in the wastewater treatment process, such as free chlorine and chloramines, absorb UV light and reduce UVT. Before UV design and equipment selection, the design team must be aware of upstream chemical use and its effects on UVT.

2.2.3.4 Industrial Effects on Ultraviolet Transmittance

Examples of industries that generate UV-absorbing wastes and byproducts include textile, printing, pulp and paper, food processing, meat and poultry, photo-developing, pharmaceutical, and chemical manufacturers. Table 5.2 lists some of the more common compounds that absorb UV light and decrease UVT. The composition of industrial wastewater varies depending on the industry; however, when industrial wastewater discharges include UV-absorbing compounds that are not well biodegraded, care must be taken to provide pretreatment; in this case, pretreatment involves the industry providing some treatment to the wastewater to target the removal of UV-absorbing compounds. An example of the effect of industrial wastewater on UV disinfection is the effect of landfill leachate on a UV disinfection system installed at the Chambers Creek Regional Wastewater Treatment Plant located in Pierce County, Washington, where landfill discharges that

TABLE 5.2 Ultraviolet-absorbing and light-scattering chemicals (Courtesy of Trojan).

Inorganic compounds/source	Organic compounds with a conjugated ring	Compounds found in water and wastewater
Bromine	Anisol Benzene	Organic dyes and coloring agents—textile industry
Chlorine	Chlorobenzene	Humic acids
Chromium	O,m, p-cresol Cyanoanthracene	Lignin sulfonates—fruit processing industry
Cobalt	O-cyclohexyl phenol	Extracts from leaves, Tannins
Iodides	Cyclohexyl phenyl ketone	Orzan S
Iron	Hexatiene	Phenolic compounds
Manganese	1-methyl-3,4 dihydronapthlene	Sunblock, PABA
Nickel	O-methylstyrene	Tea, Coffee
Sulfates	Octahydrophenathrene Phenyl propane	Carbamide compounds Pharmaceutical industry
Stannous chloride	Phenol	
Sodium thiosulphate, Photo-processing industry, Microbiology sample bottles	1-tetalone Toluene Triolefin	

represented a fraction of a percent of the plant flow caused a substantial reduction in UVT for the bulk flow. Daily average flows of leachate received at the utility caused the UVT of secondary effluent to drop by up to 9% after the landfill had been in operation for approximately 10 years. Because the utility was in the process of designing an upgraded UV system, disinfection of the lower UVT secondary effluent would have required additional UV equipment for the utility to meet regulatory dose requirements. Also, additional operation and maintenance costs (associated with energy costs and lamp maintenance) would be necessary when the leachate is discharged to the WRRF.

The utility conducted a study to confirm the effect of leachate on UV disinfection and revealed that the biological treatability, measured as the ratio of BOD to chemical oxygen demand (COD) (BOD/COD), of landfill leachate declines with the age of the landfill. At a BOD/COD of 0.4 or greater, leachate is considered to be easily degraded by biological means (Petruzzelli et al., 2007); at a BOD/COD of less than 0.4, leachate is considered to be recalcitrant to biological treatment. In comparison, municipal wastewater has a BOD/COD of about 0.5. Upon investigation, it was determined that the landfill had been in operation for more than 10 years when low UVT was measured in the secondary effluent; the biological treatment provided at the preliminary landfill treatment facility and at the utility had been previously sufficient in treating the leachate, but was no longer sufficient in recent years.

The aforementioned example illustrates that a proper understanding of industrial waste and its effect on UV disinfection is essential to determine system and/or pretreatment requirements. Some industrial wastewater may be more easily biologically degradable so that typical secondary wastewater treatment is sufficient for removal. However, industrial wastewaters that are not biologically degradable should, therefore, be the focus of preliminary treatment agreements. Bench-scale testing may be necessary to determine the ratios at which industrial wastes impart significant effects on secondary effluent UVT. Flow equalization may also be practiced to maintain a consistent dilution of industrial waste and to avoid "shocking" the UV system. Owners, operators, design engineers, and waste-producing industries must be willing to create pretreatment agreements that specifically target UV-absorbing compounds.

2.2.3.5 Sidestream Flow Effects on Ultraviolet Transmittance

Sidestream flows at WRRFs (e.g., recycling streams from solids dewatering) typically contain high COD loads that are not easily degraded in the secondary treatment process. As a result, these loads may be present in secondary effluent and reduce the UVT and, therefore, the effectiveness of UV disinfection. At the Livermore Water Reclamation Facility (Livermore, California), the effect of sidestream flows is observed in secondary effluent about 6 to 10 hours after the belt filter press filtrate pump is operated, and UVT of the secondary effluent decreases dramatically. This demonstrates how sidestream pump operation (which conveys belt filter press filtrate to the headworks) affects secondary effluent UVT. The UV design team should consider the effect of such sidestreams on secondary effluent quality and UVT. Bench-scale testing of bypass streams for UVT may be necessary to assess these effects and determine the ratios at which bypass flows affect the secondary effluent UVT. Operators may also consider flow equalization of bypass streams to maintain a consistent dilution of industrial waste and to avoid "shocking" the UV system.

2.3 Upstream Processes to Improve Water Quality

From previous discussions, it is clear that UV disinfection efficiency is influenced by effluent quality and the level of pretreatment. This underscores the importance of understanding upstream treatment processes and their effect on effluent quality, including both the value and variability of parameters, to provide an optimized system design. Preliminary testing of effluent quality will also reveal conditions that must be remedied before UV implementation, such as infrequent, but low, UV transmittance values that may occur or particle and solids loading that affects disinfection. Additionally, informed decisions about whether upstream process improvements may be needed to address these factors, can have an effect on UV system efficiency.

Upstream process control is important for improvement of downstream water quality that can affect disinfection performance. For example, Loge et al. (2002) found that, in activated sludge supernatant, the fraction of particles with associated coliform bacteria declines exponentially with increasing values of mean cell residence times in the activated sludge process. And, a study by Azimi et al. (2014) demonstrated that parameters such as activated sludge retention time, operating effluent temperature, and phosphorus deficiency may affect floc physicochemical characteristics and the UV dose required for disinfection.

Other researchers have found that implementing advanced tertiary treatments such as granular or membrane filtration resulted in better reliability of the disinfection process. Qualls et al. (1983) noted that filtration of effluent with 8-μm filters improved inactivation under low-pressure UV lamps and disinfection was significantly improved in a sand-filtered effluent compared to unfiltered secondary effluent. Jolis et al. (1999) determined that

UV following microfiltration reliably met coliform standards for wastewater reclamation in California for coliforms with a dose of 45 mJ/cm^2 and for MS2 virus with a dose of 88 mJ/cm^2. A dose of only 80 mJ/cm^2 was required when inline filtration was applied (Jolis et al., 2001).

The UV dose required to meet the various permit requirements is highly dependent on microorganism concentrations and wastewater particle interference (Linden, 1998). Providing higher levels of treatment in the secondary process, or providing tertiary filtration ahead of UV disinfection, will improve performance and reduce UV dose requirements. The UV dose required to reduce the indicator microbe concentration to the target level can be defined by the microbe's UV dose-response curve. The UV dose-response curve is a plot of the microbe's concentration on the y-axis as a function of UV dose on the x-axis. The UV dose-response curve is measured using a collimated beam experiment, as described in Chapter 2. Example collimated beam results are shown in Figures 5.4a and 5.4b from the Sunnyvale Water Pollution Control Plant in Sunnyvale, California, that uses conventional activated sludge and tertiary sand filtration; samples were collected before and after filtration to evaluate UV dose-response data to determine an appropriate level of disinfection to meet various standards.

As shown in Figures 5.4a and 5.4b, the UV dose response of microbes in wastewater is typically biphasic, with a region of first-order inactivation kinetics at low UV doses and a region of tailing or diminished inactivation at high UV doses. The first-order inactivation represents the inactivation of dispersed microbes that are not clumped, whereas the tailing region represents the inactivation particle-associated microbes that are housed within the wastewater particles and, hence, are shielded from UV light (Qualls and Johnson, 1983). Treatment operations at the WRRF, including primary settling, activated sludge, and filtration, reduce concentrations of dispersed and particle-associated microbes upstream of the UV system. For efficient UV disinfection, the upstream processes should reduce the number of particle-associated microbes so UV disinfection can meet the disinfection criteria at relatively low UV doses.

There are many different types of filters available that can effectively filter and condition the water for disinfection by UV light. These types of filters can be grouped into two categories: (1) membrane and (2) media filtration. Membrane filtration processes include microfiltration, ultrafiltration, nanofiltration, and reverse osmosis. Given that the pore size of membrane filters is much smaller than the filters used in the TSS test, all suspended solids should be removed by a membrane filter. Most bacteria are larger than the pore size (i.e., 0.1 micron) of microfiltration membranes; however, low concentrations of coliform (sometimes below detection) can

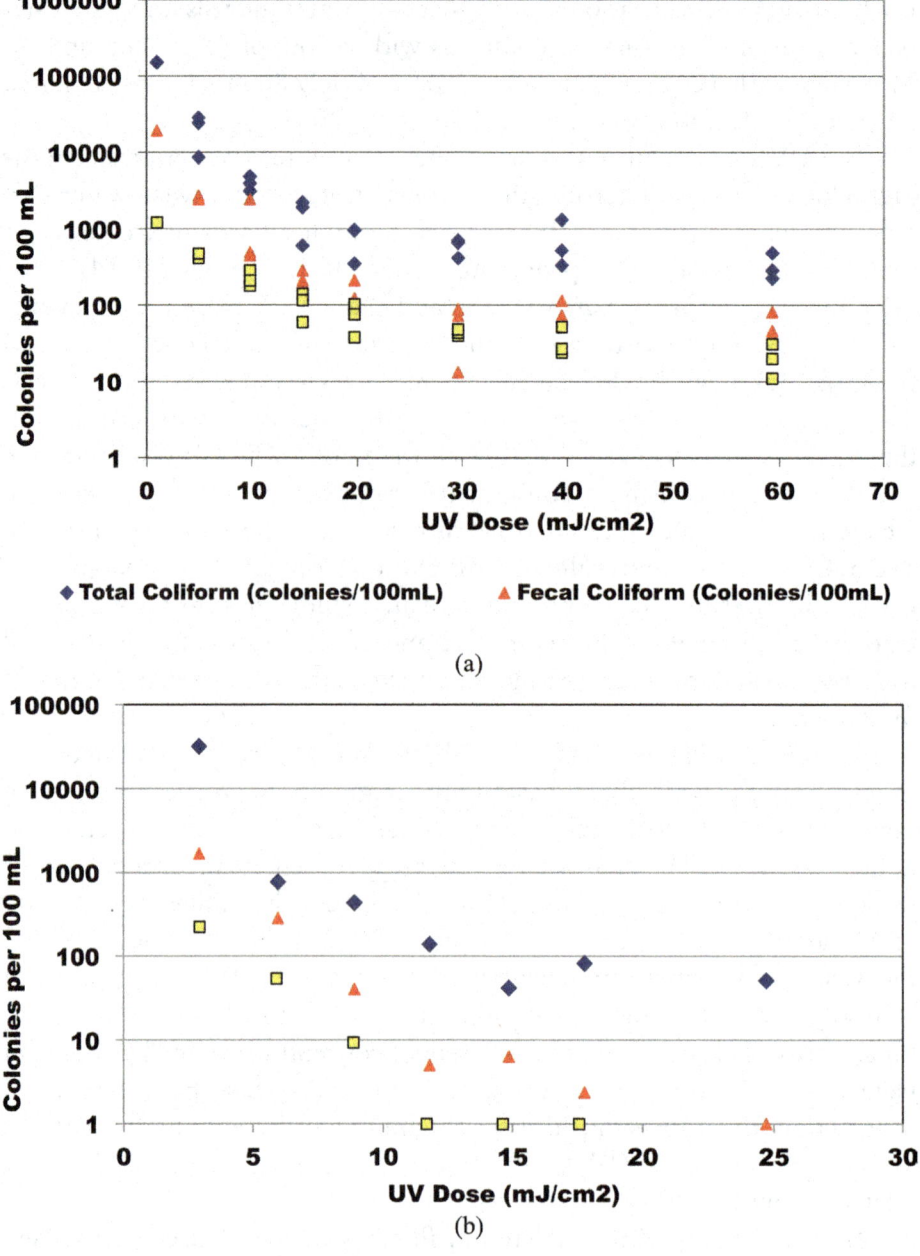

FIGURE 5.4 (a) and (b) Collimated beam data for total coliform, enterococcus, and fecal coliform vs UV dose in unfiltered and filtered effluent.

be achieved after membrane filtration, even on membrane systems that do not have membrane integrity issues. Bacteria that exist in membrane effluent will not be particle-associated, but will be free-swimming organisms that should be inactivated with a low UV dose in the range of 5 to 10 mJ/cm^2.

Media filtration includes filtration process such as granular media, automatic backwash (traveling bridge), continuous backwash, compressible media, cloth disk, and cloth and stainless steel microscreen filters. With the exception of microscreen filters, which typically have a pore size of 10 to 17 microns, most media filters do not have an absolute pore size. Without a prefiltration coagulation step (i.e., addition of a coagulant or polymer upstream of the filters), media filters typically only remove particles larger than 5 to 10 microns. To increase the particle removal rates of a tertiary filter, a prefiltration coagulation step can be included. As shown in Figure 5.5, when a prefiltration coagulation step is provided, better particle removal can be achieved.

2.4 Fouling of Lamp Sleeves, Lamp Racks, and Channels

Fouling is also an important consideration in UV system design and equipment selection. Inorganic or mineral fouling is caused by the presence of ions, such as calcium or iron, in the secondary effluent. Such ions can form deposits on the UV lamp sleeve, thereby reducing the UV energy provided to the effluent and reducing disinfection efficiency. Biological fouling is

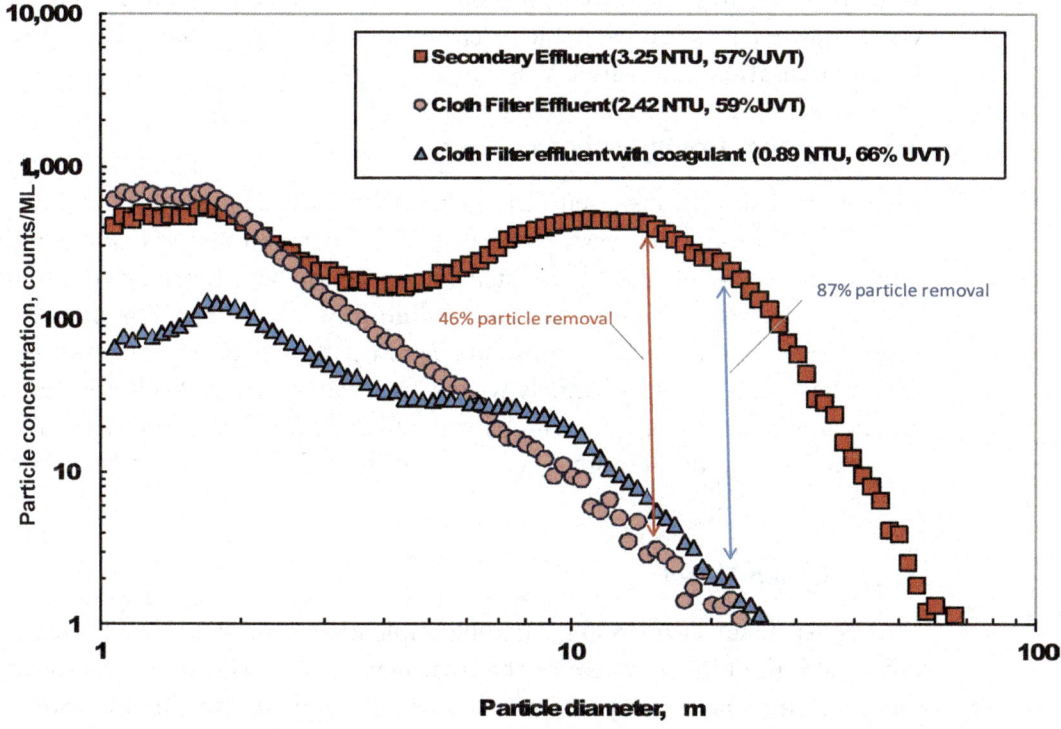

FIGURE 5.5 The effect of a prefiltration coagulation step on filtered effluent water quality.

caused by the presence of organisms (such as algae), which may develop a film on the quartz sleeve surface or on lamp racks and accumulate in channels, causing the same effect of reduced UV energy output. Fouling of UV lamp sleeves, racks, and channels must be mitigated by practicing preventive maintenance to remove deposits from the UV equipment surface. Quartz sleeve fouling factors, expressed as a percentage of the irradiance from a new and clean quartz sleeve, typically range between 0.70 and 0.95 and are highly dependent on site-specific water quality and the effectiveness of automated cleaning systems. Medium-pressure lamps, as a result of their higher intensity output and much higher lamp temperature, are more prone to fouling when compared to low-pressure and LPHO lamps. The rate at which quartz sleeves foul is also an important consideration because this information determines frequency of lamp cleaning.

3.0 REACTOR SELECTION CRITERIA

As detailed in Chapter 2, there are several lamp technologies applied in the wastewater disinfection market; in general, low-pressure, LPHO, and medium-pressure systems represent the majority of commercially available systems. Ultraviolet lamps can be configured into reactors in a number of ways, and reactors are available in closed-vessel or open-channel systems. Lamp orientation varies by UV system.

3.1 System Configurations

Ultraviolet disinfection systems can generally be divided into open-channel and closed, pressurized vessel systems, with open-channel systems being more common. While most UV systems in operation have the lamps encased in quartz sleeves and submerged in the fluid flow, there are a few noncontact systems that house UV lamps outside of the flow path. Hybrid systems, which have culvert-like channels with a free water surface both upstream and downstream, use both open-channel and closed-vessel features. The use of a particular type of system is an engineering decision and is based on site-specific conditions.

3.1.1 Closed Vessel

Ultraviolet disinfection systems can be implemented in pressurized vessels, inline with the piping system of the treatment facility. Ultraviolet lamps in these systems may be oriented parallel or perpendicular to the flow. Some systems house lamps that are configured at an angle into the flow or that extend diagonally across the reactor.

3.1.2 Open Channel

In wastewater applications, open-channel UV systems are primarily used for ease of access to the system components for maintenance and the ability to achieve low headloss, which is particularly important as a retrofit into existing chlorine contact tanks. Open-channel systems have a free water surface and operate at atmospheric pressure. Open-channel disinfection systems often consist of a number of lamp modules fit across the width of the channel. Each module has a number of lamps extending into the channel depth. Modules are organized into banks of lamps, and each system may have multiple banks in series in a channel.

3.1.3 Lamp Orientation

Open-channel systems can have lamps oriented horizontally and parallel to the water flow or vertically and positioned perpendicular to the water flow. Some vertically oriented lamps are set on a diagonal as they descend into the channel. Most closed-channel UV disinfection systems have lamps oriented across the reactor perpendicular to the flow; others may have the lamps in a mixed horizontal/vertical waffle pattern or horizontally in parallel to the flow. Alternative configurations, such as diagonal across the reactor or horizontally angled into the reactor exist as well; lamps may also be oriented in either uniform or staggered arrays.

3.1.4 Non-Submerged Ultraviolet Lamp Systems

While the majority of wastewater systems have lamps submerged in the water, non-submerged systems are also available. These systems flow water through UV-transparent tubes with lamps outside of the tubes in air, with the light penetrating into the water through the UV-transparent material.

3.1.5 Lamp Spacing

Lamp spacing required for efficient disinfection directly correlates to the water quality (UVT), the velocity of the water (related to flow), and the germicidal UV output from the lamp (related to the lamp power, and the lamp efficiency). The delivered UV dose to the water matrix will be lowest at the point equidistant between four UV lamps in a 2×2 array. Poor water quality (low UVT) will require either tighter lamp spacing or better mixing for effective disinfection. The effect of a particular lamp spacing on disinfection efficiency should be directly reflected in the UV equipment cost. This reflection will only be observed in UV systems that have undergone rigorous equipment validation and certification.

3.2 Establishing Design Criteria

Before the widespread use of bioassay validation, pilot testing to determine the design UV dose, UVT, and the UV reactor were the best means of determining design criteria for a UV disinfection system. With the advances in bioassay and other UV sizing tools, it is typically not necessary to conduct pilot tests. Bench-scale collimated beam studies, in conjunction with water quality characterization, can support design of UV systems with the use of bioassays that are available for validated reactors, resulting in expected reactor performance. However, at some utilities where high fouling potential exists or where operations staff seek hands-on experience (or if a UV reactor is preferred that needs additional validation), pilot testing can provide value. Key criteria that are important in designing a UV disinfection system are summarized as follows (additional information on some of these parameters is provided in the following sections):

- Maximum and minimum flow;
- Maximum and minimum UVT;
- Required UV dose;
- Microorganism that is used to specify the UV dose;
- End-of-lamp-life (EOLL) factor;
- Effect of the maximum quartz sleeve fouling on UV output (i.e., fouling factor);
- Maximum and minimum temperature of the wastewater; and
- Dose response curves for indicator organisms to meet the regulatory limit.

3.2.1 Flow

Ultraviolet systems can be effectively designed for any flow. Selection of a UV system should take into the account the minimum flow, maximum flow, startup flow, and average flow. The selected UV system must be able to provide the required disinfection for the full range of flows, and ideally operate efficiently over that flow range. For some systems with a high peak flow relative to the average flow, a compromise may be required between efficient reactor turndown and peak-flow disinfection robustness. Generally, closed-vessel systems are more commonly used for lower flow applications or when a small footprint is available. Recent innovations in higher-powered LPHO lamps and their inclusion in several open-channel systems allow larger flows at lower UVTs to be treated by these lamp systems. This will likely result in more applications with high flow and/or low UVT using the more efficient LPHO lamps.

3.2.2 Headloss and Water Level

Ultraviolet system performance is highly dependent on reactor hydraulics. For example, if influent fecal coliform is 100 000 most probable number (MPN)/100 mL and effluent standards require 200 MPN/100 mL, then the UV system must reduce the fecal coliform by 99.9% (3-log reduction). Assuming a reactor was designed to marginally attain this 3-log reduction, the hydraulic behavior of the reactor becomes critical. Any short-circuiting of flow (around the top, sides, etc.) can result in noncompliance; for validated systems in which reactor performance and log reduction is defined, the short-circuiting would primarily occur in open-channel systems in which construction and water-level tolerances do not meet specification.

Ultraviolet suppliers typically develop a headloss curve, as described in Chapter 3, which shows the expected headloss at varying flows for open-channel systems. There is no limit to the maximum allowable headloss in closed-vessel systems; however, the systems are validated over a flow range. This allowable headloss is a function of a combination of factors, including lamp spacing, the allowable headloss for minimizing exposure of UV lamps, and details of the validation testing, which tests the reactor performance over a set range of water levels (simulating headloss at different flowrates). For horizontal lamp systems, the acceptable water level above the top lamp will range from the top of the quartz to about one-half of the lamp spacing (and possibly slightly more based on validation testing). Too much water level over the top lamp will result in short-circuiting and too little water will result in exposure of the quartz sleeve, excessive sleeve fouling, and potential overheating of the top lamps. For vertical or inclined systems, the effects are different. By nature, the vertical system will have exposed quartz sleeves. Low water levels result in exposed lamp arc, which can cause increased fouling at the top of the sleeve and also results in wasted disinfection energy. High water levels above the lamp arc can result in short-circuiting of flow. Overall, operating the system within the validated water-level range will result in more predictable performance and reduced operation and maintenance issues. The design team should include the following hydraulic constraints in contract documents to provide a robust disinfection system:

- The maximum water surface elevation at which the UV system can operate must be based on third-party reactor validation testing;
- The minimum water surface elevation at the UV banks that the UV system can operate must be based on third-party reactor validation testing. The quartz sleeves of the UV banks for horizontal systems shall be almost completely submerged at this minimum water surface elevation;

- For all level control devices (weir, mechanical, or motorized level control gate), the specified minimum water surface elevation shall be maintained in the open channel;
- Maximum acceptable normalized lamp velocity (peak flow rate divided by the number of lamps in a bank) must be specified by the UV supplier and shall not exceed the maximum value proven effective during third-party validation testing; and
- Minimum acceptable normalized lamp velocity shall be specified by the UV supplier and shall not be less than the minimum value proven effective during third-party validation testing.

3.2.3 Influent and Effluent Water Quality

Any selected UV disinfection system must be able to continuously provide the minimum UV design dose at the design flow and design UVT. Water quality design inputs include several (but not always all) of the following: UVT, TSS, water temperature, and influent and required effluent concentration of the target organism. Often, collimated beam dose response data are provided to the design team to better size the UV system. However, for new facilities, where effluent is not available for testing, design criteria can be based on information from similar facilities; many UV equipment manufacturers maintain databases of information on organism concentrations from various processes. In either case, the UV disinfection system can be based on performance requirements that are delineated by third-party bioassay.

3.3 Design Dose and Dose Control Strategies

Ultraviolet systems are most often specified in one of two ways. First, they can be specified to meet a particular effluent concentration of indicator organisms under worst-case effluent quality conditions and maximum fouling of the quartz sleeves at the end of lamp life. Secondly, a UV system can be specified such that it must produce a certain UV dose (fluence), with the worst-case effluent and maximum quartz sleeve fouling at the end of lamp life. An example of the first situation is when a UV system must meet a concentration of fecal coliforms of 200/100 mL. Specifying dose (fluence) as reduction equivalent dose (RED) (MS2 or otherwise) can be used for nonreuse applications and for water reuse. An example of the second situation is when the UV system must meet an MS2 RED (fluence) of 40 mJ/cm^2. Specifications can have one or both of these conditions, but only one can be used unless the required UV dose (fluence) was obtained from UV dose- (fluence)-response curves with the indicator organism.

In most cases, the UV dose (fluence) that is specified is far above what is required to meet the regulatory limit for the indicator organisms. A required RED can also be specified based on the sensitivity of the indicator organism instead of the challenge organism when the equipment specified has a UV dose (fluence). The bioassay can be used for these situations and all of the required information should be provided by a validation report. In the case of specified indicator microorganisms, it is necessary to have UV dose- (fluence)-response curves (collimated beam tests) for the indicator microorganisms when the effluent is at its worst-case condition; challenge organisms should have the same or similar UV sensitivity. In the case with the required UV dose (fluence) or RED, the dose must be specified with respect to one of the challenge organisms (e.g., an MS2 RED of 30 mJ/cm^2 or a T1 RED of 15 mJ/cm^2). If a wastewater has abnormal levels of iron or high hardness, then pilot testing may be advisable.

There are many UV equipment systems that are available in the municipal market with varying instrumentation and controls capabilities. All significant UV manufacturers include an advanced system control center to control the UV disinfection system at each bank or reactor. Criteria for selection of UV system controls are typically based on the utility's budget and other requirements. However, UV systems should have minimum monitoring and control capabilities, as described in the following sections.

3.3.1 System Monitoring

It is advantageous for larger facilities to monitor parameters including UVT, flow, water level, and UV intensity. These data are generally used to operate each train or channel individually to deliver the specified dose. Ultraviolet transmittance monitoring can be performed either immediately upstream or downstream of the UV system. Water level should be monitored in each open channel, and UV intensity should be monitored for each UV reactor or bank. Smaller facilities typically do not need to monitor UVT and water level. In smaller facilities, the UVT can be entered into the control system manually after analyzing a grab sample with a bench-top unit. If a level control weir is installed, there is no need to monitor water level because it will only fluctuate a few inches. Ultraviolet disinfection system instrumentation and control features can add significant cost to the overall UV project, but operational savings can be realized by taking advantage of these features.

3.3.2 Dose Delivery Strategies

In addition to providing all the information that is required to specify a UV system for a WRRF so it will provide the proper UV dose (fluence) under the worst-case conditions, the validation report should also describe the algorithm for calculating UV dose (fluence). The dose equation included in the third-party reactor validation test report must be incorporated to the control algorithm to continuously calculate the delivered dose of the system, automatically vary the lamp power and control the system as required to minimize energy use, and deliver the specified dose at all times. There are a number of control strategies that can be used and these are summarized in the following subsections.

3.3.2.1 Dose Pacing Based on Flow, Ultraviolet Transmittance, and Power Setting

The dose-pacing program should use the equation developed in the third-party reactor validation. Ultraviolet intensity sensors are not integrated to the calculation of dose, and there are a number of systems that operate well without properly calibrated and accurate sensors. Under this control method, the UV sensors provide intensity monitoring and sleeve fouling and lamp-aging factors must be manually included in the control formula. In the event that the system has calibrated and accurate UV sensors, but they are not included in the main control philosophy, the system can be programmed to have those sensors monitoring the combined effect of lamp aging and sleeve fouling, and thus can be used as a tool to help signal when maintenance is required.

3.3.2.2 Dose as a Function of Flow, Ultraviolet Transmittance, and Ultraviolet Intensity

As in all dose-pacing methods, the control program should use the equation that was developed in the third-party reactor validation testing. For this method, the sensor value is used as the primary control parameter (along with flow and UVT) to continuously monitor the combined effect of lamp power for variable output UV systems, lamp aging and sleeve fouling, and effluent quality.

The UV sensor-based dose equation is emerging as a preferred control method because it allows for meeting the minimum required UV dose with the least amount of energy while still generating an alarm when the delivered UV dose is less than the minimum acceptable UV dose setpoint.

4.0 ULTRAVIOLET SYSTEM SIZING EXAMPLE

The UV dose (fluence) that must be used to disinfect a wastewater is either specified by a regulatory agency or determined by the design engineer. If the design dose is not specified by a regulatory authority, then the design engineer may determine the dose using site-specific information and collimated beam experiments, as described in Chapter 2. If the facility represents a new construction, the design dose may be estimated based on experience from other similar facilities, as noted in Section 3.2.3. Once the design criteria have been established, the third-party bioassay can be used as a tool to determine the level of disinfection that is provided under given conditions at a WRRF. It is important to note that the bioassay can only provide information about the performance of a system that was within the validation conditions that were used during the third-party testing (sometimes called the *validation envelope*). As an example, the bioassay cannot predict performance of the system if suspended solids are present or if the quartz sleeves are not transmitting the minimum amount of UV light they were tested with. Thus, it is important to produce dose- (fluence)-response curves of the indicator organisms through collimated beam tests, for the worst-case wastewater, to determine the effect of suspended solids and UVT.

Design parameters for the "Anywhere, U.S.A." UV project are detailed in Table 5.3. Other factors that affect sizing of UV systems that are not addressed in the bioassay are the EOLL factor and the fouling factor (F_F). The EOLL factor refers to the the number of lamp hours when the UV output of the lamp has reached a specified percentage of the new lamp output, and is considered to be the point at which the lamp has reached the end of its life. The F_F varies between different types of automatic cleaning systems from different UV manufacturers. Organic and inorganic fouling

TABLE 5.3 Ultraviolet design and system configuration for Anywhere WRRF, U.S.A.

Key design parameters	Design dose	Enterococci RED of 35 mJ/cm^2, minimum
	Design method	NWRI Guidelines, UVDGM & IUVA Uniform Protocol
	Design (peak) flow	16 MGD (2,524 m^3/hr)
	Average flow	6.5 MGD (1,025 m^3/hr)
	UV transmittance	65%, minimum
	Total suspended solids	<20 mg/L
	Enterococci permit limit	35 cfu/100 mL, based on a 30-day geometric mean
		89 cfu/100 mL, daily maximum
System configuration	Number of channels	2
	Number of banks per channel	2

accumulates on the quartz sleeve and reduces the effective UV irradiance. The quartz sleeve fouling factor is expressed as a percentage of the irradiance from a new quartz sleeve. Both of these factors should be validated by an independent third party so that they can be confidently used in UV system sizing. For this example, the EOLL is 0.85 and the F_F is 0.90.

As shown in Table 5.3, the indicator organism for this design is Enterococcus; a summary of several collimated beam tests from the facility is presented in Figure 5.6. Based on several collimated beam experiments, the engineering team determined that an Enterococci RED of 35 mJ/cm² was a conservative dose for the facility to comply with the enterococci permit limit of 35 cfu/100 mL, based on a 30-day geometric mean and an 89-cfu/100-mL daily maximum.

The dose-per-log (D_L) for the enterococci data is approximately 8 mJ/cm²/log I; this was determined by analyzing the steep portion of the dose-response curve, non-particle-associated enterococci. The linear portion of this curve shows a dose of approximately 24 mJ/cm² and a Log I of 3, which equates to a D_L of 8 mJ/cm²/Log I (24 mJ/cm² divided by 3). Using the equation shown below from the bioassay exemplar presented in Chapter 3, all of the variables are known except for Q (i.e., the flow per lamp in gallons per minute [liters per second]).

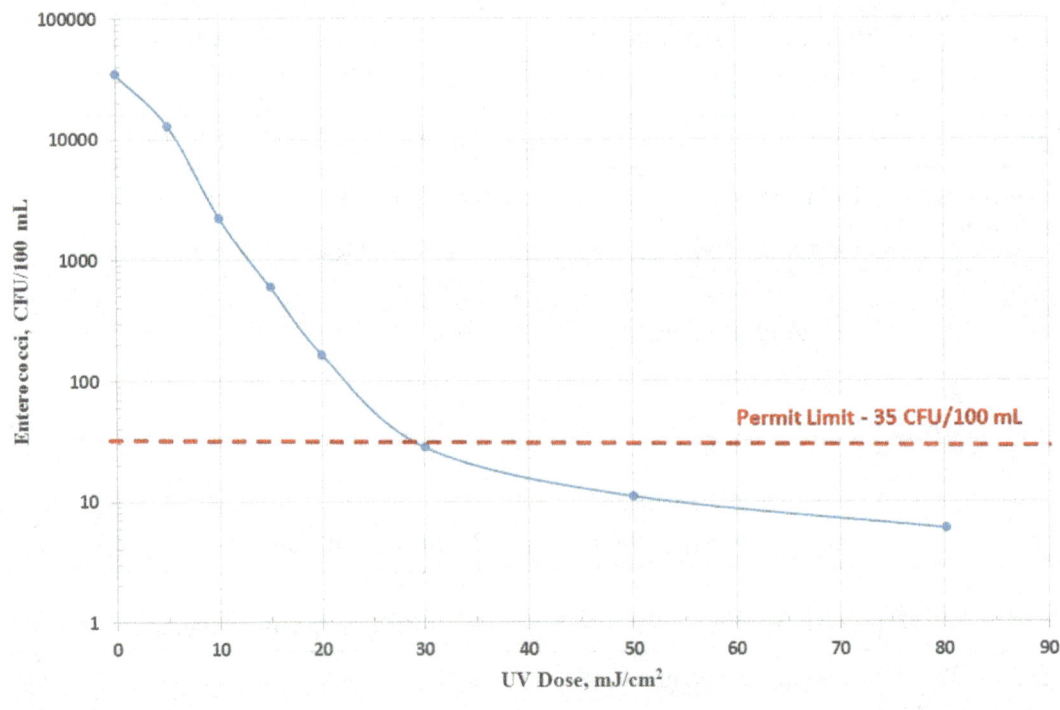

FIGURE 5.6 Anywhere WRRF secondary effluent collimated beam test results.

$$\text{Log I} = (10^{0.1612} \times A_{254}^{-4.188}) \times \left[\frac{S/S_0}{Q \times \text{Sensitivity}} \right]^{(-0.1528 \times Ln)(A_{254}) + 0.6669} \times \text{Banks}^{0.9454}$$

Based on the design parameters in Table 5.3, known variables are defined as follows:

Log I = dose/sensitivity = 35 mJ/cm² ÷ 8 mJ/cm²/Log I = 4.375;
Sensitivity = 8 mJ/cm²/Log I;
A_{254} = −log (UVT/100) = −log (65/100) = 0.1871;
S/S_0 = EOLL × F_F = 0.85 × 0.90 = 0.765; and
Banks = number of banks in series = 2, as specified in the design parameters.

Solving for Q (flow/lamp) provides a flow rate of 118 gpm (0.447 m³/min) per lamp. Based on the design parameters and system configuration, this represents the following lamp count per bank:

Q_{max} = 16 mgd = 11 111.1 gpm (42.06 m³/min),
Number of channels = 2,
$Q_{max/channel}$ = 5555.6 gpm (21.03 m³/min), and
Number of lamps per bank = $Q_{max/channel}$ ÷ Q = 5555.6 ÷ 118 = 47.1.

Based on the equipment manufacturer's largest lamp rack (8 lamps per rack), this represents 5.9 lamp racks. Rounding up to the next whole number, each bank will have a total of 6 lamp racks. Therefore, the UV system will have the following system configuration based on the design parameters and facility-specific collimated beam data:

Number of channels:	2
Number of banks per channel:	2
Number of lamp racks per bank:	6
Number of lamps per lamp rack:	8
Total number of lamps:	192

This system has been matched to the specific requirements of the facility. The aforementioned system provides an Enterococci RED of 35.6 mJ/cm². It is important that this example does not provide any process redundancy, and additional equipment will need to be added based on owner requirements. Additional discussion on process redundancy is provided in Chapter 6.

5.0 REFERENCES

Amoah, K.; Craik, S.; Smith, D. W.; Belosevic, M. (2005) Inactivation of *Cryptosporidium* Oocysts and Giardia Cysts by Ultraviolet Light in the Presence of Natural Particulate Matter. *J. Water Supply Res. Technol. Aqua*, 54, 165–178.

Azimi, Y.; Allen, D. G.; Seto, P.; Farnood, R. (2014) Effect of Activated Sludge Retention Time, Operating Temperature, and Influent Phosphorus Deficiency on Floc Physicochemical Characteristics and UV Disinfection. *Ind. Eng. Chem. Res.*, 53 (31), 12485–12493.

Beltran, N. A.; Jimenez, B. E. (2008) Faecal Coliforms, Faecal Enterococci, Salmonella Typhi and Acanthamoeba spp. UV inactivation in Three Different Biological Effluents. *Water SA*, 34, 261–269.

Bianchini, R.; Calucci, L.; Lubello, C.; Pinzino, C. (2002) Intermediate Free Radicals in the Oxidation of Wastewaters. *Res. Chem. Intermediates*, 28 (2-3), 247–256.

Bohrerova, Z.; Linden, K. G. (2006) Ultraviolet and Chlorine Disinfection of Mycobacterium in Wastewater: Effect of Aggregation. *Water Environ. Res.*, 78, 565–571.

Bohrerova, Z.; Mamane, H.; Ducoste, J. J.; Linden, K. G. (2006) Comparative Inactivation of Bacillus subtilis Spores and MS-2 Coliphage in a UV Reactor: Implications for Validation. *J. Environ. Eng.*, 132 (12), 1554–1561.

Bolton, J. R.; Stefan, M. I.; Cushing, R. S.; Mackey, E. (2001) The Importance of Water Absorbance/Transmittance on the Efficiency of Ultraviolet Disinfection Reactors. *Proceedings of the First International Congress on Ultraviolet Technologies*; Washington, D.C., June.

Buffle, J.; Leppard, G. G. (1995) Characterization of Aquatic Colloids and Macromolecules. 1. Structure and Behavior of Colloidal Material. *Environ. Sci. Technol.*, 29 (9), 2169–2175.

Camper, A. K.; Buls, J.; Goodrum, L. (2001) Effects of UV Disinfection on Humic Substances and Biofilms. *Proceedings of AWWA Water Quality Technology Conference*; Nashville, Tennessee.

Dai, X.; Hozalski, M. R. (2002) Effect of NOM and Biofilm on the Removal of *Cryptosporidium parvum* Oocysts in Rapid Filters. *Water Res.*, 36, 3523–3532.

Darby, J.; Emerick, R.; Loge, F.; Tchobanoglous, G. (1999) The Effect of Upstream Treatment Processes on UV Disinfection Performance; Project 96-CTS-3; Water Environment Research Foundation: Alexandria, Virginia.

Droppo, I. G. (2001) Rethinking What Constitutes Suspended Sediments. *Hydrol. Process.,* **15** (9), 1551–1564.

Emerick, R. W.; Loge, F. J.; Ginn, T.; Darby, J. L. (2000) Modeling the Inactivation of Particle-Associated Coliform b=Bacteria. *Water Environ. Res.,* **72** (4), 432–438.

Frimmel, F. H. (1998) Characterization of Natural Organic Matter as Major Constituents in Aquatic Systems. *J. Contam. Hydrol.,* **35**, 201–216.

Gehr, R.; Nicell, J. (1996) Pilot Studies and Assessment of Downstream Effects of UV and Ozone Disinfection of a Physicochemical Wastewater. *Water Qual. Res. Can.,* **31**, 263–281.

Gehr, R.; Wagner, M.; Veerasubramanian, P.; Payment, P. (2003) Disinfection Efficiency of Peracetic Acid, UV and Ozone after Enhanced Primary Treatment of Municipal Wastewater. *Water Res.,* **37**, 4573–4586.

Gerba, C. P.; McLeod, J. S. (1976) Effect of Sediments on the Survival of Escherichia coli in Marine Waters. *Appl. Environ. Microbiol.,* **32** (1), 114–120.

Gregory, J. (1997) The Density of Particle Aggregates. *Water Sci. Technol.,* **36** (4), 1–13.

Guenther, E. A.; Johnson, K. S.; Coale, K. H. (2001) Direct Ultraviolet Spectrophotometric Determination of Total Sulfide and Iodide in Natural Waters. *Anal. Chem.,* **73** (14), 3481–3487.

Guo, M.; Huang, J.; Hu, H.; Liu, W. (2011) Growth and Repair Potential of Three Species of Bacteria in Reclaimed Wastewater after UV Disinfection. *Biomedical Environ. Sci.,* **24** (4), 400–407.

Ho, C.-F. H.; Pitt, P.; Mamais, D.; Chiu, C.; Jolis, D. (1998) Evaluation of UV Disinfection Systems for Large-Scale Secondary Effluent. *Water Environ. Res.,* **70**, 1142–1150.

Jagger, J. (1967) *Introduction to Research in Ultraviolet Photobiology;* Prentice-Hall: Eaglewood Cliffs, New Jersey.

Jolis, D.; Lam, C.; Pitt, P. (2001) Particle Effects on Ultraviolet Disinfection of Coliform Bacteria in Recycled Water. *Water Environ. Res.,* **73** (2), 233–236.

Jolis, D.; Hirano, R.; Pitt, P. (1999) Tertiary Treatment Using Microfiltration and UV Disinfection for Water Reclamation. *Water Environ. Res.,* **71**, 224–231.

Keen, O. S.; Love, N. G.; Linden, K. G. (2012) The Role of Effluent Nitrate in Trace Organic Chemical Oxidation during UV Disinfection. *Water Res.,* **46** (16), 5224–5234.

Lester, Y.; Sharpless, C. M.; Mamane, H.; Linden, K. G. (2013) Production of Photo-Oxidants by Dissolved Organic Matter During UV Water Treatment. *Environ. Sci. Technol.*, **47** (20), 11726–11733.

Linden, K. G. (1998) UV Disinfection for Wastewater: State of the Technology. *Civ. Eng.*, March.

Liu, G.; Slawson, R. M.; Huck, P. M. (2007) Impact of Flocculated Particles on Low Pressure UV Inactivation of E-coli in Drinking Water. *J. Water Supply: Res. Technol.*, **56**, 153–162.

Loge, F. J.; Bourgeous, K.; Emerick, R. W.; Darby, J. L. (2001) Variations in Wastewater Quality Influencing UV Disinfection Performance: Relative Impact on Filtration. *J. Environ. Eng.*, **127** (9), 832–837.

Loge, F. J.; Emerick, R. W.; Ginn, T. R.; Darby, J. L. (2002) Association of Coliform Bacteria with Wastewater Particles: Impact of Operational Parameters of the Activated Sludge Process. *Water Res.*, **36** (1), 41–48.

Loge, F. J.; Emerick, R. W.; Thompson, D. E.; Nelson, D. C.; Darby, J. L. (1999) Factors Influencing Ultraviolet Disinfection Performance Part I: Light Penetration to Wastewater Particles. *Water Environ. Res.*, **71** (3), 377–381.

Mackey, E. D.; Cushing, R. S.; Wright, H. B. (2001) Effect of Water Quality on UV Disinfection of Drinking Water. *Proceedings of the First International Ultraviolet Association Congress;* Washington, D.C., June 14–16.

Mackey, E. D.; Wright, H.; Hargy, T.; Fonseca, A. C.; Cabaj, A. (2004) Reducing UV Design Safety Factors, Optimization of UV Reactor Validation. *Proceedings of the American Water Works Association Water Quality Technology Conference*; San Antonio, Texas, Nov 14–18.

Madge, B. A.; Jensen, J. N. (2006) Ultraviolet Disinfection of Fecal Coliform in Municipal Wastewater: Effects of Particle Size. *Water Environ. Res.*, **78**, 294–304.

Mamane, H. (2008) Impact of Particles on UV Disinfection of Water and Wastewater Effluents: A Review. *Rev. Chem. Eng.*, **24** (2-3), 67–157.

Mamane-Gravetz, H.; Linden, K. (2004) UV Disinfection of Indigenous Aerobic Spores: Implications for UV Reactor Validation with Unfiltered Waters. *Water Res.*, **38** (12), 2898–2906.

Metcalf & Eddy/AECOM (2014) *Wastewater Engineering: Treatment and Resource Recovery*, 5th ed.; McGraw-Hill: New York.

Metz, D. H.; Reynolds, K.; Meyer, M.; Dionysiou, D. D. (2011) The Effect of UV/H_2O_2 Treatment on Biofilm Formation Potential. *Water Res.*, **45**, 497–508.

Nelson, K. L. (2000) Ultraviolet Light Disinfection of Wastewater Stabilization Pond Effluents. *Water Sci. Technol.*, **42**, 165–170.

Nourmoradi, H.; Nikaeen, M.; Stensvold, C. R.; Mirhendi, H. (2012) Ultraviolet Irradiation: An Effective Inactivation Method of *Aspergillus* spp. in Water for the Control of Waterborne Nosocomial Aspergillosis. *Water Res.*, **46** (18), 5935–5940.

Pereira, V. J.; Weinberg, H. S.; Linden, K. G.; Singer, P. C. (2007) UV Degradation Kinetics and Modeling of Pharmaceutical Compounds in Laboratory Grade and Surface Water via Direct and Indirect Photolysis at 254 nm. *Environ. Sci. Technol.*, **41** (5), 1682–1688.

Petruzzelli, D.; Boghetich, G.; Petrella, M.; Dell'erba, A.; L'abbate, P.; Sanarica, S.; Miraglia, M. (2007) Pre-Treatment of Industrial Landfill Leachate by Fenton's Oxidation. *Global NEST J.*, **9** (1), 51–56.

Qualls, R. G.; Flynn, M. P.; Johnson, J. D. (1983) The Role of Suspended Particles in Ultraviolet Disinfection. *J. Water Pollut. Control Fed.*, **55** (10), 1280–1285.

Qualls, R. G.; Johnson, J. D. (1983) Bioassay and Dose Measurement in UV Disinfection. *Appl. Environ. Microbiol.*, **45** (3), 872–877.

Scheible, K. O. (1987) Development of a Rationally Based Design Protocol for the Ultraviolet Light Disinfection Process. *J. Water Pollut. Control Fed.*, **59** (1), 25–31.

Severin, B. F.; Suidan, M. T.; Engelbrecht, R. S. (1983) Kinetic Modeling of UV Disinfection of Water. *Water Res.*, **17**, 1669–1678.

Sharpless, C. M.; Seibold, D. A.; Linden, K. G. (2003) Nitrate Photosensitized Degradation of Atrazine during UV Water Treatment. *Aquat. Sci.*, **65** (4), 359–366.

Shubert, W.; Gunthert, F. W. (2001) Particle Size Distribution in Effluent of Trickling Filters and in Humus Tanks. *Water Res.*, **35**, 3993–3997.

Templeton, M. R.; Andrews, R. C.; Hofmann, R. (2005) Inactivation of Particle-Associated Viral Surrogates by Ultraviolet Light. *Water Res.*, **39**, 3487–3500.

U.S. Environmental Protection Agency (2006) *Ultraviolet Disinfection Guidance Manual for the Final Long Term 2 Enhanced Surface Water Treatment Rule*; EPA-815/R-06-007; U.S. Environmental Protection Agency: Washington, D.C.

Urbain, V.; Block, J. C.; Manem, J. (1993) Bioflocculation in Activated Sludge: An Analytical Approach. *Water Res.*, **27** (5), 829–838.

Wait, I. W.; Blatchley, E. R. (2010) Model of Radiation Transmittance by Inorganic Fouling on UV Reactor Lamp Sleeves. *Water Environ. Res.*, **82** (11), 2272–2278.

Wilkinson, K. J.; Negre, J. C.; Buffle, J. (1997) Coagulation of Colloidal Material in Surface Waters: The Role of Natural Organic Matter. *J. Contam. Hydrol.*, **26**, 229–243.

6

Equipment Selection, Facility Design, and Project Delivery

Katherine (Kati) Y. Bell, Ph.D., P.E., BCEE, and Joshua E. Goldman, Ph.D.

1.0	INTRODUCTION	130	2.4.2 Constructability and Maintenance of Facility Operations during Construction	141
2.0	LIFE CYCLE COST ANALYSIS	130		
	2.1 Project Capital Costs	131		
	2.1.1 Ultraviolet Equipment Capital	131	2.4.3 Operation and Maintenance Considerations	142
	2.1.2 Construction Costs	132	2.4.4 Warranties, Service, and Manufacturer Reliability	142
	2.2 Operation and Maintenance Costs	133		
	2.2.1 Power Consumption and System Efficiency	133	2.4.5 Reference Installations and Other Considerations	143
	2.2.2 Replacement Parts	134	3.0 FACILITY DESIGN CONSIDERATIONS	143
	2.2.2.1 Lamps and Sleeves	134	3.1 Site and System Hydraulics	143
	2.2.2.2 Ballasts and Drivers	135	3.1.1 Available Head	144
	2.2.3 Cleaning Components	136	3.1.2 Open-Channel Versus Closed-Channel Systems	144
	2.2.4 Intensity Sensors and Ultraviolet Transmittance Analyzers	136	3.1.3 Flow Splitting, Flow Distribution, and Flow Control	145
	2.2.5 Operations and Maintenance Labor	136	3.1.4 Flow Measurement	145
	2.3 Calculating Life Cycle Costs	137	3.1.5 Level Control	146
	2.4 Non-Cost Considerations	140	3.2 Power Requirements and Power Redundancy	146
	2.4.1 Headloss Effects	140	3.3 Process Redundancy	147

3.4	Ultraviolet System Layout	148	3.4.3 Sleeve Cleaning Methods and Ancillary Facilities	151
	3.4.1 Lifting Devices	149		
	3.4.2 Ultraviolet System Ballast Cabinets, Control Panels, and System Instrumentation	150	4.0 REFERENCES	152

1.0 INTRODUCTION

There have been significant advances in UV equipment system development since being introduced to the wastewater market over three decades ago. Whereas the fundamental operating principles of the available equipment systems are generally similar, the physical differences in equipment systems offered by various UV manufacturers are significant enough to affect the procurement, design, and costs of these systems. These are important considerations in project delivery. For example, an equipment procurement process may require preparation of multiple designs to allow bidding of multiple UV equipment systems. In this case, the final equipment selection is often left up to the general contractor, with the decision typically based on lowest capital cost. While this method ensures a competitive capital price for the owner, it does not always take into account differences in long-term operation and maintenance (O&M) costs, unique UV equipment features, or support services provided by the competing UV manufacturers. There are other equipment procurement options that typically include the following: sole-source of equipment with design around the selected equipment; design around a single supplier, allowing others to bid as long as they prepare a revised design; or competitive preselection of equipment with the design being performed around the selected equipment. Chapter 7 contains an additional discussion of procurement methods along with their advantages and disadvantages. The following sections provide an overview of information commonly considered for procurement of UV equipment, with a specific example on calculation of total project cost. The sections also cover significant design considerations in development of bid and procurement documents and the evaluation of UV bids through life cycle cost analysis.

2.0 LIFE CYCLE COST ANALYSIS

There are diverse equipment configurations that correspond to the same basis of design (e.g., reduction equivalent dose, flow, UV transmittance [UVT], and effluent bacteria concentration) because UV system layout, lamp count, and other ancillary equipment requirements can vary substantially from manufacturer to manufacturer. It is critical to be able to compare different equipment on a life cycle basis to understand how these system configurations affect the project to select the most cost-effective project delivery method, equipment procurement method, and overall best UV disinfection solution. It is important to note that while the equipment capital costs for a UV system are significant, the costs of power and replacement parts, such as lamps and ballasts, can be as much as the initial equipment capital costs when considered over the project life (typically 10 to 20 years). Thus, understanding the full cost of ownership for a UV system is a critical part of the equipment selection process.

Selection of UV equipment is often driven by costs; however, many procurement methods also provide a means of considering other technical criteria. This section covers the development of life cycle costs and provides an example from a recent project. Comparison between UV systems can be challenging because there are numerous configurations available to meet the same disinfection treatment objectives. In addition, the same UV equipment package may vary in cost based on the project delivery and equipment procurement method used. Decisions about these methods are multifaceted and are both site- and owner-specific. Once the project delivery method (further described in Chapter 7) and equipment procurement method are defined, it is possible to obtain UV equipment proposals such that life cycle costs can be developed to allow comparison of various UV equipment systems. Life cycle costs should include capital equipment and construction costs, replacements-part costs (based on the warranty period), chemical and labor costs, power costs, and several other ancillary factors. Finally, non-cost factors may also be considered when making final decisions about the best UV system for a specific application.

2.1 Project Capital Costs

The heart of any UV project is the equipment; and, whereas this cost is often only a fraction of the total construction cost, the cost must be captured, along with an understanding of the UV suppliers' scope of supply to make accurate estimates of both construction and life cycle costs. In addition to equipment costs, the total capital costs also include any structural, civil, electrical instrumentation and controls, and other support systems necessary for project implementation.

2.1.1 Ultraviolet Equipment Capital

To obtain the most accurate UV equipment costs, a detailed request for proposal should be prepared, including a complete set of proposal forms, contracting forms, technical specifications, warranties, and drawings showing the equipment layout and scope of supply limits of the UV manufacturer. Whereas the criteria for selection of the UV manufacturer can be based on capital cost, it is also important to obtain replacement part costs to aid the owner in developing O&M budgets and life cycle costs. Although it is not within the scope of this chapter to provide a UV equipment system specification, the key components of a well-developed specification are described.

After the design criteria for the system have been developed (including design dose and system redundancy requirements, as described in Chapter 5), the system scope of supply should be clearly defined. Generally, UV systems are specified using performance requirements that are based on third-party validated bioassay dose, a specific surrogate organism, maximum influent and target effluent organism concentrations, wastewater flowrate, wastewater UVT, maximum allowable aging and fouling factors, suspended solids concentrations, and other relevant factors.

A number of approaches can be used for obtaining proposal costs, ranging from specifying only the UV equipment to requiring the UV manufacturer also provide all required support equipment; this includes items such as the step-down transformer, influent and/or effluent isolation gates, level control devices, instruments, and lifting devices. The key advantage of the latter method of specification development is placing all of the equipment that is controlled by the UV equipment supplier under a single point of responsibility; however, the drawback is that the UV equipment supplier may mark up those costs, resulting in a higher equipment system cost. The specification package should also include any special provisions for UV equipment acceptance and performance testing (including laboratory analyses), headloss and electrical harmonics testing, training, and equipment system and system component warrantees.

Although a generic specification can be used to obtain preliminary or conceptual equipment costs, it may be difficult to compare proposal costs from different UV equipment suppliers because packages can have a vastly different scope of supply, making the packages difficult to compare. Therefore, the best results are obtained when a full specification that includes requirements for construction materials, costs for additional site support, and lamp guarantees is developed before obtaining equipment proposals from UV suppliers.

2.1.2 Construction Costs

Construction costs will vary based on the UV system configuration, footprint, and ancillary requirements, as described in Section 3.0. Construction cost estimates can be developed at any point in a project, but the level of confidence and utility of the cost estimate depends on the accuracy of the cost estimate and the level of detail that is used to develop it. The American Association of Cost Estimators International has developed a cost-estimate classification system that maps the phases and stages of project cost estimating into five classes of cost estimates that provide varying levels of accuracy (AACE, 2012). For example, Class 1 cost estimates are based on 65 to 100% design information and are appropriate for project bidding; methods that could be used to obtain these conceptual costs include detailed unit costs with detailed takeoffs. The level of accuracy of these projects is -3 to $+10\%$. As the class of cost estimates increases, the level of accuracy decreases because estimates are based on less detailed design information. The lowest level of estimates are Class 5, which are based on 0 to 2% design information, and are appropriate for concept screening; these estimates are typically based on pricing guides and cost curves, judgment, and cost analogy from similar projects. The level of accuracy of these projects is -20 to $+50\%$.

Construction costs should include installation of elements included in the UV specification as well as integration with existing instrumentation and control systems, installation of pipes and valves, interstage pumping, temporary construction facilities, housing, sitework and excavation, subsurface considerations, standby power, contingencies, and engineering inspection costs.

2.2 Operation and Maintenance Costs

Initial equipment and construction capital are only part of the costs that should be considered with respect to budgeting and selecting a UV disinfection system. There are annual O&M costs that include cost of electrical consumption and demand charges, replacement parts, and O&M labor. This discussion will provide an overview of significant O&M costs that should be considered in a life cycle cost analysis for a UV project.

2.2.1 *Power Consumption and System Efficiency*

Power efficiencies vary widely by lamp and ballast type, and power costs will be site-specific. Thus, in some cases, life cycle costs of various systems may balance out when the entire cost of ownership, or life cycle cost, is considered. For example, low-pressure systems may be more efficient from a power consumption standpoint than medium-pressure systems, but medium-pressure systems may have lower capital and construction costs; thus, it is important to consider power consumption costs and demand charges in any life cycle analysis. The power consumption as a function of design dose and water quality characteristics can be estimated by information presented in the validation report provided by the manufacturer. Generally, however, for typical wastewater characteristics, power consumption rates of 0.4 to 1.0 kW/ML·d (1.5 to 4 kW/mgd) can be expected; for medium-pressure systems, power consumption rates are on the range of 2 to 3 times that of low-pressure systems, depending on equipment system and effluent water quality.

2.2.2 Replacement Parts

Considering a typical project life cycle of 10 to 20 years, the cost of replacement parts can be a substantial factor in the overall life cycle cost of a UV project. Parts that require replacement regularly include UV lamps and ballasts/drivers and cleaning components such as wipers. There are other parts that may need to be replaced because of breakage, such as quartz sleeves. A spare parts inventory of these system components that includes a percentage of the total parts installed (typically about 10%) should be kept on-site. The actual life of a component is a function of many variables, including operating conditions, maintenance practices, the quality of construction materials, and fabrication practices. Consequently, estimating the actual life of every component can be a challenge, and an adequate inventory of critical spare parts should be maintained to ensure reliable and consistent performance of the UV equipment.

Additionally, UV equipment components have both a design life and a guaranteed life. The design life is the expected duration of operation and the guaranteed life incorporates the risk, assumed by the manufacturer, to account for the uncertainties associated with the quality of materials used, production, and operating conditions. Generally, guarantees are conditional and are valid under specified operating conditions.

2.2.2.1 Lamps and Sleeves

Ultraviolet lamp technology has advanced to the point where suppliers have an understanding of how germicidal lamp output changes with time. When a new UV lamp is installed, the output is typically high, and as the lamp ages with use, the output will typically decline. Thus, before lamps fail, they decline in performance over time (Figure 6.1). This is accounted for in UV equipment sizing by a design end-of-lamp-life (EOLL) factor. Ultraviolet lamp manufacturers have characterization information on the lamps they provide; this information is used in determining the amount of time that the lamp would be guaranteed to meet a specified output.

The cost of lamps over the life of a UV project is not insignificant, and lamp manufacturers or UV equipment suppliers will typically provide a lamp guarantee that provides lamp replacement if the output of a lamp does not meet the specified EOLL within a lamp warranty period. While most UV vendors provide a lamp warranty, there can be wide variations in the details of these warranties. For example, some suppliers will prorate the replacement lamp cost after some number of operating hours; other suppliers will only

Chapter 6 ■ Equipment Selection, Facility Design, and Project Delivery

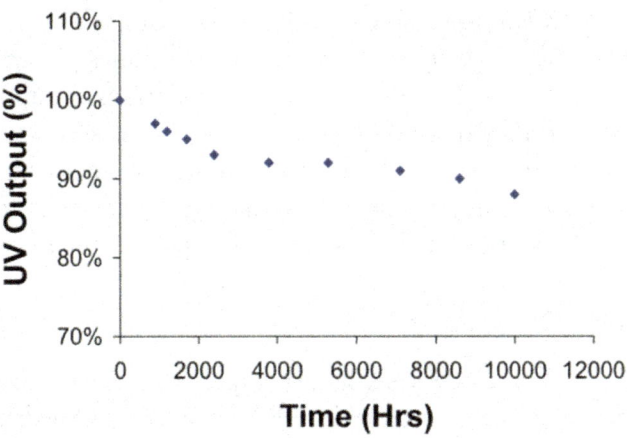

FIGURE 6.1 Germicidal output as a function of lamp aging for a typical low-pressure high-output mercury lamp, from U.S. EPA (2006).

provide a warranty for a period from the date of shipment, regardless of the number of operating hours that the lamp has been operated.

Other considerations in evaluating lamp replacement costs are whether a lamp supplier will provide lamps for a fixed cost (or some fixed cost tied to a cost index) over the life cycle of the project, which can be specified during procurement, or whether the supplier will pass on higher costs for replacement parts after the system is put into operation. Thus, it is important to understand these provisions and how they affect the annual costs for lamp replacement in terms of both operations and the overall effect on project life cycle cost estimates.

While UV sleeves are anticipated to last longer than UV lamps, lamp sleeves can degrade over time because of solarization (i.e., darkening of the quartz) and internal sleeve fouling, resulting in cloudiness and loss of UV transmittance. Abrasion of the sleeve surface during handling or mechanical cleaning may also contribute to the loss of UV transmittance through the quartz sleeves. And, in wastewater where iron salts are used, consideration for inorganic fouling may be important in determining the frequency of replacing quartz sleeves. There can also be attrition of these quartz sleeves because of breakage or in some site-specific applications. Thus, sleeve replacement should also be considered in a life cycle analysis.

2.2.2.2 Ballasts and Drivers

The lamp ballast or electronic driver is an electrical device that provides the proper voltage and current required to initiate and maintain the gas discharge within the UV lamp. Power supplies and ballasts are available in many different configurations and are tailored to a unique lamp type and application. Both electronic and magnetic ballasts are used in wastewater equipment, and each has specific advantages and disadvantages, although the majority of equipment that is used in water resource recovery facilities (WRRFs) uses electronic ballasts or drivers. The life of electronic ballasts is substantially longer than the lamps that they drive; however, because they are power devices and can generate heat, and it is important to consider environmental conditions under which they operate in terms of estimating replacement costs. Ballasts and drivers can be warranted for a specific life span by the manufacturer.

2.2.3 Cleaning Components

Online mechanical cleaning systems use wipers that are attached to either electric motors or pneumatic or hydraulic piston drives. Mechanical wipers may consist of stainless steel brush collars or Teflon rings that move along the lamp sleeve. A life cycle analysis may account for replacement of wipers and/or cleaning solutions, when required. The schedule for cleaning solution refilling ranges from monthly to annually. The frequency of replacement of wipers will vary depending on the specific system and effluent water quality, but can range from yearly to every 3 years. The rate of chemical cleaning also can range significantly depending on the site-specific water quality and degree and frequency of fouling; generally, chemical cleaning can be anticipated to be conducted monthly to annually.

2.2.4 Intensity Sensors and Ultraviolet Transmittance Analyzers

Ultraviolet intensity sensors and UVT analyzers are important in controlling dose delivery in UV disinfection systems. They are highlighted here because an owner should budget for maintaining, calibrating, and replacing these instruments as needed. The expected life of these instruments will vary from manufacturer to manufacturer and as a function of water quality but, generally, an owner should anticipate replacing UV-intensity sensors one to two times during the life of a project; UVT, if well maintained, may last the entire project period.

2.2.5 Operations and Maintenance Labor

The design, operation, and maintenance needs for UV disinfection differ from those of traditional chemical disinfectants used in wastewater applications. The extent of fouling will determine the amount of maintenance (labor and parts) that is needed on a routine basis and will affect the overall maintenance costs. The labor costs for any project can be estimated by consulting with the UV manufacturer to obtain estimates of time per operation or maintenance task, considering the frequency of these tasks, which is dependent on the specific system and effluent water quality. To accurately reflect labor costs, loaded labor costs should be used that reflect the various pay and fringe rates for a facility.

2.3 Calculating Life Cycle Costs

Construction cost estimates are an important part of the life cycle analysis. Depending on the point in the project at which the equipment selection occurs, different levels of cost estimates are appropriate. There are a number of methods to conduct a life cycle analysis, but, at a minimum, life cycle costs should include power, labor costs for O&M, and replacement parts costs for lamps, sleeves, drivers, and other equipment that will likely need replacement during the project life cycle; all of these factors will vary by UV manufacturer. While these factors will vary from system to system, the approaches to how these factors are evaluated are presented in this section.

An example project is presented in this section to demonstrate how the factors listed in the previous section are incorporated to a life cycle analysis. The project described is a tertiary WRRF with a 700-L/s (16-mgd) average flow and a peak hourly flow of 128 ML/d (33 mgd). The project specifications established design doses for both MS2 and T1 to meet monthly geometric mean *Escherichia coli* of 126 colony forming units (cfu)/100 mL, with no single sample exceeding 941 cfu/100 mL. Redundancy requirements included either one redundant bank per channel or one redundant channel at peak hourly flow. A specification was developed for UV equipment and all ancillary equipment that would be required for operations of the UV system; this included transformers, inlet isolation gates, effluent finger weirs, and performance testing for equipment acceptance, including bacterial disinfection performance, electrical consumption, electrical harmonics, and headloss. Using the equipment proposals provided by four manufacturers, Class 3 construction cost estimates were developed in support of preselection of equipment that would serve as the basis of design for the traditional design/bid/build project. Table 6.1 presents a summary of the project capital cost estimates showing the breakdown of significant cost-estimate line items.

As noted in the previous sections, there are a number of replacement parts that have an effect on the annual operations costs of a UV disinfection system. While the actual design and guaranteed life of various UV components are specific to each manufacturer's system, typical design and guarantee lives are summarized in Table 6.2.

Each UV system manufacturer provided data on power consumption at design average daily flow conditions as well as replacement costs for its lamps, ballasts, wipers, quartz sleeves, and cleaning chemicals, where applicable. Calculation of annual costs for this example product, summarized in Table 6.3, included the following assumptions:

- Treated flow for all 20 years of operation would be 62 ML/d (16 mgd);
- The significant cost components included power consumption, lamp replacement, ballast replacement, and quartz sleeve replacement;

TABLE 6.1 Ultraviolet disinfection system comparative capital costs.

Cost factor	Manufacturer A ($)	Manufacturer B ($)	Manufacturer C ($)	Manufacturer D ($)
Base system cost				
UV disinfection system equipment	855,360	517,814	759,000	443,499
Spare parts and accessories	74,300	22,086	Included	Included
18-month warranty	44,430	10,530	Included	7,085
Manufacturer's services[a]	26,360	35,750	Included	20,205
Adders supplied by UV manufacturer				
Module cleaning station	11,560	10,100	n/a[b]	n/a
Submersible pump for cleaning tank	1,130	1,285	n/a	n/a
PLC software licenses	Included	6,010	Included	10,000
Underwriters Laborties listing for panels	Included	7,750	Included	Included
Deducts supplied by UV manufacturer				
Replace PLC with CompactLogix PLC	−34,900	−3,846	−5,100	0
Provide NEMA 3R Transformers	−2,300	−6,600	−3,200	−7,785
Additional project requirements				
Lifting equipment	Included	69,000	Included	Included
Concrete structure	231,000	104,000	302,000	224,000
Pre-engineered metal canopy	69,000	77,000	108,000	76,000
Effluent isolation gates	20,000	n/a	n/a	n/a
Electrical/I&C	323,985	212,720	290,175	193,251
Total direct costs	**1,657,125**	**1,074,045**	**1,459,175**	**974,040**
Permits	8,286	5,370	7,296	4,870
Sales tax	89,523	52,372	72,105	42,132
Builder's risk	15,090	9,730	13,225	8,780
General liability	30,180	19,460	26,450	17,560
Bonds and insurance	45,270	29,190	39,675	26,340
Subtotal before Ohead & Profit	**1,845,474**	**1,190,167**	**1,617,926**	**1,073,723**
General conditions	184,547	119,017	161,793	107,372
Contractor's overhead and profit	184,547	119,017	161,793	107,372
Subtotal with OH&P	2,214,569	1,428,200	1,941,511	1,288,467
Construction contingency	553,642	357,050	485,378	322,117
Total cost at today's dollars	2,768,211	1,785,251	2,426,889	1,610,584
Escalation to midpoint of construction	249,348	160,807	218,603	145,074
TOTAL CAPITAL COST	**3,018,000**	**1,946,000**	**2,645,000**	**1,756,000**

[a]Includes training, harmonic studies, and testing.
[b]n/a = not applicable to UV manufacturer's design.

TABLE 6.2 Typical design and guaranteed lives of significant UV components.

Component	Component design life[a]	Guaranteed life[b]
Low-pressure lamps	12 000–15 000 hours	10 000–12 000 hours
Medium-pressure lamps	10 000 hours	5000–8000 hours
Quartz sleeves	8–10 years	1–3 years
UV-intensity sensors	3–10 years	1 year
UVT analyzer	5–10 years	1 year
Cleaning wipers	3–5 years	1–3 years
Ballasts	10–15 years	1–5 years

[a]Expected duration of operation
[b]Accounts for variability of material quality, production, and operating conditions

TABLE 6.3 Estimated annual costs at 62-ML/d (16-mgd) average daily flow.

Cost factor	Manufacturer A	Manufacturer B	Manufacturer C	Manufacturer D
Power consumption				
Annual power consumption, kWh	462 000	256 000	387 000	326 000
Annual power cost @ $0.095/kWh	$43,900	$24,300	$36,800	$31,000
Lamp replacement				
Lamps replaced per year	53	53	23	31
Cost per replacement lamp	$220	$150	$450	$185
Annual cost for replacement lamps	$11,700	$8,000	$10,400	$5,700
Ballast replacement				
Ballasts replaced per year	15	8	2	3
Cost per replacement ballast	$400	$285	$880	$400
Annual cost for replacement ballasts	$6,000	$2,300	$1,800	$1,200
Quartz sleeve replacement				
Quartz sleeves replaced per year	6	5	3	3
Cost per replacement sleeve	$90	$65	$158	$164
Annual cost for quartz sleeves	$500	$300	$500	$500
Total annual O&M cost	$62,000	$35,000	$50,000	$38,000

- Quantities of lamps and ballasts replaced are based on the anticipated component life;
- Approximately 2% of quartz sleeves will be replaced each year because of attrition; and
- The unit cost for power was $0.095/kWh.

Using the information developed in Tables 6.1 and 6.3, a net present cost can be calculated. The following cost considerations should be included in a net present cost calculation (a summary of the analysis is provided in Table 6.4):

- Determine the project life cycle (10 to 20 years is the typical range of a planning period for UV systems);
- The discount rate and inflation rate will depend on project-specific funding conditions;
- Establish the year capital costs will be incurred for construction and the year the system will be fully operational; and
- A conservative estimate may include replacement of all quartz sleeves at year 7 to 10, depending on the system.

2.4 Non-Cost Considerations

There are non-cost factors that owners may consider when selecting a UV disinfection system. Although these criteria are unique to each owner, there are a number of factors that should be considered during a technical evaluation. Non-cost criteria are often scored qualitatively, with scoring based on a scale of most desirable to least desirable; these qualitative scores can be weighted and added to a cost score to support the selection process. Some of the factors that are often considered during a non-cost evaluation are summarized in the following subsections.

2.4.1 Headloss Effects

Many UV projects represent retrofits to existing facilities, where the hydraulic gradeline is already set; as a result, it is often desirable to minimize the

TABLE 6.4 Summary of net present cost for example system.

Cost factor	Manufacturer A	Manufacturer B	Manufacturer C	Manufacturer D
Total NPC of capital costs	$3,018,000	$1,946,000	$2,645,000	$1,756,000
Total NPC of annual O&M costs	$963,000	$541,000	$766,000	$596,000
Total NPC	$3.98M	$2.49M	$3.41M	$2.36M

headloss of the UV system design. The headloss of a UV system is unique to each equipment system, but also depends on the influent and effluent isolation and the level control strategy used in the design. As a result, systems with low headloss may provide more flexibility during design.

2.4.2 Constructability and Maintenance of Facility Operations during Construction

Ultraviolet channels and reactors may be placed outdoors or indoors depending on site conditions, climate, and cost considerations. Outdoor facilities typically have a canopy structure to protect equipment and maintenance staff from the climatic elements when servicing the equipment; indoor UV systems require a more expensive building structure that typically includes space for the electrical components of the UV system. Regardless of location, UV channels are typically covered with fiberglass or aluminum panels to prevent UV light from exiting the channel and to block external light to reduce algae growth. The panels also create a safe work area and protect the equipment from objects entering the channel and causing damage.

Other issues include consideration of construction tolerances, which is important for open-channel UV systems. These systems are typically installed in concrete basins with a free water surface and appropriate flow uniformity (e.g., diffuser plate) and water level control devices, and require tight channel width and height tolerances (as little as 0.25 in. [6 mm]) where the UV lamp banks are situated. Some contractors may have difficulty meeting such tight channel tolerances and thus the work requires close inspection and potential subsequent modification. Generally, constructability typically depends on site-specific conditions, although closed-vessel UV systems are typically easier to construct compared to open-channel UV systems.

Because many UV projects are retrofits to existing facilities, the constructability and ease of expansion may be considerations in selection of the equipment system. Expandability may be required as a result of changes in operating conditions, such as an increase in flow or a drop in UVT because of the addition of some industrial customers. For open-channel systems, a spare channel can be constructed, but not fitted, with UV equipment or the channel walls can be designed with "break-out" features to allow for increased future capacity. For closed-vessel systems, blind flanges can be installed for future installations or reactors can be added in series. Maintenance of facility operations during construction should be considered during design so that the contractor can provide needed equipment or materials during construction.

2.4.3 Operation and Maintenance Considerations

General O&M requirements that should be considered when evaluating various systems should include ease-of-lamp and ballast replacement and cleaning system requirements, including replacement of wipers. While additional discussion is provided in Chapter 8, it is important to note that that there may be significantly different requirements among various systems; it is also important to consider these factors during selection of a system.

The total number of lamps and the frequency of replacement drive most of the maintenance time in a well-designed system. The latest low-pressure high-output systems use high-power lamps that allow for larger flows to be treated with less equipment, minimizing the time for O&M. Lamp replacement for many closed-vessel systems can be done without taking a reactor offline. Open-channel systems require mechanical means to remove the lamps from the channel and, for large, open-channel systems, removal and replacement of lamps using a crane can be a time-consuming task. Reactor accessibility is also a consideration in reactor selection; generally, it is simpler to access equipment in open-channel systems. Wiper replacement, reactor inspection, and reactor cleaning may require substantial effort in closed-vessel systems compared to an open-channel system.

2.4.4 Warranties, Service, and Manufacturer Reliability

System and system component warranties can vary widely, in addition to the service and reliability offered by various manufacturers. Thus, it is important that these factors are specified so that the owner has confidence in the ability to obtain responsive support after startup. The specification should have equipment and performance warranty requirements clearly defined to protect the owner from costs related to defects and low performance of the UV system. For example, if the system fails to meet specified performance criteria during the warranty period and the UV supplier is unable to implement corrective actions, the UV supplier could be responsible for removal and replacement of the nonconforming system. Payments for replacement or corrections not paid by a UV supplier as part of a procurement contract must be paid from the UV supplier's performance bond.

Quality of equipment service is often a function of both the manufacturer and its local or regional representative. The relative availability of spare parts is also important and can affect the delivery times and availability of spare parts, which may have an effect on the number of spare parts that should be inventoried at the facility. Additionally, some manufacturers use lamps that are specifically manufactured for that supplier and other UV manufacturers indicate that use of aftermarket lamps would void the system warranty.

2.4.5 Reference Installations and Other Considerations

If an owner is not experienced with UV disinfection, then it is recommended that he or she check manufacturer references from similar facilities. This would enable operations staff to ask questions about the experiences with similar installations. Reviewing the O&M manuals provided by vendors for the installation can inform the operations staff about the type of support that is available for a given system; a well-written manual will facilitate operational understanding of the system performance and is typically indicative of the ease of working long term with a manufacturer. Finally, some funding mechanisms may require that equipment systems meet "buy American" requirements, which may be important in selecting the system.

3.0 FACILITY DESIGN CONSIDERATIONS

Once the project delivery method and equipment procurement method have been determined, it is then possible to develop design and bid documents, including final specifications and drawings necessary for the project. As noted, in some instances UV equipment will be preselected or prequalified and a single design may be performed; alternatively, but less commonly, more than one design (multiple sets of drawings) may be required to meet an owner's procurement policies. Regardless of the UV equipment system that is selected, there are critical design considerations for facility design that are common to all systems. The most important factors include hydraulics and power supply, but there may also be ancillary requirements such as site civil design, equipment lifting systems, adequate cooling of electrical enclosures, and system enclosures (buildings or canopies).

3.1 Site and System Hydraulics

Ultraviolet dose is a function of UV intensity and residence time. Residence times of flow through a UV disinfection system are on the order of tenths of a second for medium-pressure lamp systems and seconds for low-pressure lamp systems. In theory, optimal dose delivery is obtained with plug flow hydraulics through a UV reactor, but, in practice, UV reactors do not have ideal hydrodynamics. Non-ideal flow conditions are accounted for by using a bioassay-based approach to sizing, but excessive turbulence inside a UV reactor can result in lower doses than those predicted by the bioassay. For example, turbulence and eddies can form in the wake behind lamp sleeves oriented perpendicularly to flow. Some manufacturers recommend use of baffles to improve hydrodynamics in the reactor, but these kinds of improvements to reactor hydraulics are often obtained at the expense of headloss. Inlet and outlet conditions can also significantly affect reactor hydrodynamics and UV dose delivery. Changes in flow direction at inlets and outlets can result in short-circuiting, eddies, and dead zones within the reactor. Regardless of whether the UV system is an open-channel or closed-vessel reactor, use of straight inlet configurations with gradual changes in cross-sectional area will help create flow conditions for optimal dose delivery. Additionally, there are often minimum channel distances or piping configuration requirements that are established during bioassay validation that must be considered during design if the UV equipment will be guaranteed to deliver the design dose.

3.1.1 Available Head

Generally, headloss will be greater in closed-vessel systems than in open-channel systems because of system geometry. Headloss across an open channel will also vary between different UV systems, and hydraulic information should be included in the specification.

3.1.2 Open-Channel Versus Closed-Channel Systems

Generally, most wastewater treatment facilities are designed for gravity flow through the facility and, as a result, there is often limited head available, which is why open-channel designs are more commonly used for wastewater applications than closed-vessel reactors. Additionally, open-channel systems may provide a greater ease of access for system maintenance. However, there are a number of examples where closed-vessel UV systems have been used, particularly where effluent pumping is used for discharge.

In open-channel systems, it is critical to design the system such that there is consistency of water depth to provide proper submergence of the UV equipment. Any rapid variations in flow through the UV system that can affect the disinfection efficiency must also be considered; for example, consideration should be given to systems where tertiary filter backwash water is drawn from upstream of the UV system. Control of water level through the UV system is critical to providing effective disinfection and, if there are rapid changes in water level, then mitigation of these fluctuations should be addressed through hydraulic controls and programming logic.

While closed-vessel reactors are less common for wastewater applications, they are used; these systems are more commonly installed in filtered effluent applications with low solids concentrations and the need to access equipment for maintenance is less. With increasing use of membrane bioreactors in wastewater treatment, closed-vessel reactors are becoming more prevalent where flow is already in a pressurized pipe. When considering hydraulic controls for closed-vessel reactors, it is important to ensure that the reactor vessel is flowing full at all times.

3.1.3 Flow Splitting, Flow Distribution, and Flow Control

Regardless of whether an open-channel, gravity, or a closed-vessel reactor system is used, even flow splitting between channels or trains is critical to disinfection performance. Each channel or train (or in the case of closed-vessel reactors, piping) should be sized and configured to provide approximately equal headloss through each UV reactor train over the range of flowrates for which the facility is designed. Additionally, the flow through each reactor must conform to the operating conditions that are appropriate for each reactor. In addition to providing equal flow distribution among channels or trains, it is important to provide a uniform flow distribution within each channel. Thus, when entrance conditions to the channel are not ideal, a baffle plate may be used; however, use of such devices comes at the cost of headloss.

Two approaches for flow distribution and control are generally used. The first is active flow control and distribution, in which a dedicated flow meter and modulating control valve or gate are installed for each UV train. Other methods to achieve an active flow split include a splitter box, Parshall flumes, cutthroat flumes, and downward-opening slide gates (with electronic actuators). For closed-vessel systems, a modulating control valve on each UV train is frequently used. The second method is passive flow distribution, typically achieved through a series of baffles. For the passive approach, equal flow split may be monitored with flow meters.

Poor flow split is a frequent cause of underperformance of UV systems. And, in addition to good hydraulic practices, tools such as computational fluid dynamics modeling, described in Chapter 4, may also be used to make the most of flow splitting.

3.1.4 Flow Measurement

The method of flow measurement selected should be based on the variability in facility flow, the type of flow split used, and whether the installation will be on an open- or closed-channel system. The selection of the flowrate measurement method should be at the discretion of the owner and the design engineer based on experience and professional judgment. Generally, each UV channel or train should have a flow meter or level sensor to confirm that the flow and water level in the channel or reactor train is operating within the validated flow range. This is particularly important because, in many UV systems, flow is used as a primary control in the dose delivery algorithm. However, the industry trend is moving toward dose-paced systems controlled by UV-intensity sensors located on the UV lamp banks.

3.1.5 Level Control

As previously noted, different UV systems and configurations result in different scenarios for controlling level. Hydraulic control for closed-vessel reactors is generally straightforward; the design engineer needs to provide that the reactor vessel is flowing full at all times, although there may be an upper limit on the pressure that can be tolerated within the reactor vessels. In open-channel systems, however, there are various options for level control, and selection of the most appropriate method depends on size of flow and headloss. Options for managing level include use of motorized weir gates that can be automatically adjusted to control level as flow changes; on the other hand, where headloss is small, it may be desirable to minimize maintenance of motors and use fixed weirs. However, use of fixed weirs may be limited in some applications when the weir length to overcome larger headloss becomes too long to be cost-effective. There are also submerged flap gates or counterweighted level control gates that have been used for water level control; however, use of these devices should consider recirculation to keep up with leakage through flap gates or flap valves that prevent backflow if they will be operating with a high turn-down ratio or will see low flowrates. Submerged flap gates and counterweighted level control gates frequently cannot provide appropriate level control over the flow ranges that are typical in most municipal WRRFs.

In gravity flow applications, it is also important to include provisions for addressing unforeseen flows that could result in a facility if design flows are exceeded. Thus, in many open-channel designs, it is desirable to include either a bypass of the UV equipment system, an overflow that can be used to protect equipment from flooding, or a flood pumping system

3.2 Power Requirements and Power Redundancy

The electrical power configuration should account for the power requirements of the selected equipment, disinfection objectives, and power quality issues, if applicable. The proper supply voltage and total load requirements should be coordinated with the UV manufacturer, considering the available power supply. Power requirements for each UV system component may differ. For example, the UV reactor may require 3-phase, 480-v service, while the online UVT analyzer may need single-phase, 120-v service. Excluding pumping and blowers, the electrical load from UV reactors will typically be one of the larger loads at the facility.

Because of the varying nature of UV reactor nonlinear loads, current and voltage harmonic distortion can be induced. This can cause problems, including overheating of power supply components, and can affect other critical systems, such as variable frequency drives, programmable logic controllers, and computers. Thus, to address the potential for the equipment to induce harmonic distortion, the UV facility design and UV equipment should conform to the Institute of Electrical and Electronic Engineers (1993) Standard 519-1992, *Recommended Practices and Requirements for Harmonic Control in Electrical Power Systems.*

The continuous operation of the UV reactor is highly dependent on the power supply and its quality. If the power reliability requirements and, consequently, the disinfection objectives cannot be met by relying solely on the commercial power supply, the use of backup power, power conditioning equipment, or both may be necessary. Power reliability requirements are often specified by regulatory agencies in discharge permits. A simple backup power supply (e.g., generator) may be sufficient if power quality issues are infrequent. If an existing backup power supply is in place, its load capacity should be assessed to determine whether it can accept the additional load associated with the UV facility. The time necessary for switching from the primary power supply to a backup power supply should be considered; this gap in power can be addressed using an uninterruptible power supply, which can also provide continuous power in the event of voltage sag or power interruption. The battery capacity should be sized to be large enough to supply power to all connected equipment until a generator starts. Uninterruptable power supply systems can either be online or offline. Power conditioning can be accomplished using active series compensators to protect electrical equipment against momentary voltage sags and interruptions. These devices are well suited for instantaneous sags and interruptions; however, they cannot correct sustained sags or interruptions.

3.3   Process Redundancy

The issue of process redundancy can be complicated. While many regulatory authorities have specific design criteria with respect to process redundancy for disinfection, this is not universally the case. For example, in the United States, some states require compliance with reliability criteria set forth by the U.S. Environmental Protection Agency for wastewater treatment design (U.S. EPA, 1969), which indicates that, for disinfection, there shall be a sufficient number of units of a size such that with the largest flow capacity unit out of service, the remaining units shall have a design flow capacity of at least 50% of the total design flow to that unit operation. Unfortunately, this provides little clarity and there can be a great deal of variability in interpretation of this requirement. Ultraviolet disinfection equipment systems are modular in nature; in open-channel systems, individual modules comprise a bank, and several banks may be installed in series in a channel; closed-vessel reactors may be arranged in series in trains. Thus, several approaches have been used successfully for providing process redundancy. While the appropriate level of process redundancy depends on the specific project and regulatory requirements, some of these approaches are summarized as follows:

- Use of one redundant train/channel ($n+1$ channels/trains);
- Use of one redundant bank (reactor) per channel (train); and
- Providing peak flow capacity in the channel with one redundant (standby) bank, which could be put into operation as needed.

Other alternatives for providing process redundancy include maintaining an alternate disinfection system as backup, such as keeping existing chlorine contact basins (in retrofit applications) in operating condition for backup disinfection, or installation of a peracetic acid system that can be used in the contact volume that is available in the UV structure. However, advances in UV disinfection technology have been so significant that the desire to eliminate the maintenance of a separate secondary backup disinfection system often results in removal or demolition of these facilities in favor of providing simple equipment and electrical redundancy.

3.4 Ultraviolet System Layout

The UV facility layout is dictated by site constraints and the UV equipment constraints described in previous sections. Additionally, the following items should be considered when developing the UV reactor and facility configuration to develop the UV facility footprint in the planning phase:

- Number, capacity, dimensions, and configuration of the UV reactors (including redundancy and connecting piping or structures);
- Vertical or horizontal orientation of the UV reactor;
- Maximum allowable separation distance between the UV reactors and electrical controls if distance limitations apply;
- Provide adequate distance between adjacent reactors to afford access for maintenance tasks (e.g., lamp replacement) and spare parts storage;
- Configuration of the inlet/outlet piping or channels before and after each UV reactor based on validated hydraulic conditions and UV manufacturer recommendations; and
- Space for electrical equipment, including control panels, transformers, ballasts, backup generators, and possible uninterruptible power supplies.

Once these significant considerations for layout have been evaluated, a decision must be made regarding whether the installation will be indoors or outdoors. If the UV equipment is provided outdoors, consideration should be given to algae management, which is always a consideration for effluent that has residual concentrations of nitrogen and phosphorus; UV channels that are installed in direct sunlight should be covered. Wastewater equipment is often provided for outdoor installations with provisions for appropriate National Electrical Manufacturers Asssociation ratings; however, it is often desirable to provide either a canopy or to house the equipment indoors for a variety of reasons. For example, there can be heat buildup inside the UV electrical enclosures that can shorten the life of ballasts and other electrical components. Ballast enclosures and electrical enclosures can be located indoors in an air-conditioned room or the enclosures themselves can be air-conditioned if located outdoors. Heat load data for all ballast enclosures and electrical enclosures should be obtained from the manufacturer so the design can include features for appropriate cooling of these enclosures. Sunlight can degrade gaskets around switches, wire races, and other sealed fixtures, which can lead to gasket failure and leaks. Additionally, facility O&M staff may prefer to work under the protection from heat, cold, or precipitation. Some states, such as New Hampshire, require that some UV equipment be located indoors. Design of the facilities should include provisions for adequate work areas for maintenance activities such as laydown areas for UV reactors or modules, if equipment requires removal from service for maintenance. For example, when open-channel systems are designed, safety provisions for preventing falls into the channel should be included when modules have been removed from service. If a vertical equipment system is provided, a safe step system with fall protection may be is needed. Additionally, if acid dip tanks are included, they must be protected from infiltration and incorporate spill containment.

3.4.1 Lifting Devices

While many UV systems have integrated lifting systems as part of the equipment package, there are a number of systems that require lifting systems to remove modules from service. If the equipment will be constructed without a cover or building, it is recommended to install a robust hoisting system that is rated for outdoor use. Typical systems include jib cranes or, for small systems, a davit crane may suffice. When the UV system is located in a building or under a canopy, an overhead lifting device may be installed in the structure if the structure is rated for the additional load. Additionally, while there are systems that are manually operated, many O&M staff prefer electrically operated lifting equipment. Finally, it is important to consider where modules will be set when they are removed from service; a clear, level, area that has sufficient space for access to all sides of the equipment is important for safety and ease of maintenance activities.

3.4.2 Ultraviolet System Ballast Cabinets, Control Panels, and System Instrumentation

Regardless of which lamp type is used, the lamps are driven by ballasts that have a required range of temperatures under which they can operate. While some manufacturers of open-channel systems take advantage of the ability to cool ballasts by installing them in the space above the channel, many systems have ballasts installed in separate cabinets with ventilation fans to maintain the ballast temperature below the maximum specified limit. Design engineers should consider the ballast requirements, obtain heat load calculations, and provide a design that includes an appropriate ballast cooling system and allows for sufficient access by operations staff to perform inspection and maintenance as recommended by the manufacturer.

Most UV equipment systems have controls that can either be run locally or integrated to the plant supervisory control and data acquisition system. Typically, local control is through a human machine interface (HMI) panel that is often simply a screen on the control panel or ballast cabinet. Many wastewater installations are located outdoors where sunlight can cause difficulty in reading the HMI screen; provisions should be included to protect the HMI screen from sunlight so it can be viewed in daylight. This also can extend the life of the HMI by protecting it from the elements; for example, in cold environments, the HMI may also need to be protected from freezing temperatures with a heater element.

The control of UV systems is dependent on signals from various instruments for operation to meet disinfection doses as determined by system validation. While there is a range of operational strategies, from flow control at a set UVT to more sophisticated methods that are based on dose control using intensity sensors, the following instrumentation may be included as part of a UV equipment system: flow meters and/or channel level sensor, UVT analyzer, and UV-intensity sensors. In some wastewater equipment systems, UV sensors not only measure the UV intensity at a point within the UV reactor, but also are used to control UV dose delivery. The sensor responds to changes in lamp output because of lamp power settings, lamp aging, lamp sleeve aging, and lamp sleeve fouling. Depending on sensor position, UV-intensity sensors may also respond to changes in UVT of the water being treated. Ultraviolet-intensity sensors comprise optical components, a photodetector, an amplifier, housing, and an electrical connector. Systems that control dose using UV-intensity sensors automatically account for the degradation in UV intensity that occurs because of component aging and/or damage and adjust the system settings to achieve the correct dose. In contrast, dose delivery in flow-paced systems depends on fouling and aging factors, which are programmed into the PLC logic. Thus, flow-paced systems always apply the UV dose assuming the end-of-lamp life and fouled quartz sleeves, under all conditions.

As previously discussed, UVT is an important parameter in determining UV dose delivery. Ultraviolet transmittance analyzers are essential if UVT is part of the dose-monitoring strategy; however, there are systems that use a manually set UVT and the analyzer is not necessarily part of the system. If UVT is not part of the dose-monitoring strategy, analyzers may be provided to monitor water quality and help to diagnose operational problems. Several commercial UV reactors use the measurement of UVT to calculate UV dose in the reactor and, if necessary, change lamp output or the number of energized lamps to maintain appropriate UV dose delivery. There are a number of commercial online UVT analyzers available; one type of analyzer calculates UVT by measuring the UV intensity at various distances from a lamp. This type of analyzer is mounted external to the UV reactor and a sidestream of water passes through a cavity containing a low-pressure lamp, with UV sensors located at various distances from the lamp; the difference in sensor readings is used to calculate UVT. Recent advances in these types of UVT analyzers include use of a light-emitting diode as the UV light source, which extends the life of the analyzer. The other type of online UVT analyzer is a flow-through spectrophotometer that uses a monochromatic UV light source at 253.7 nm. The instrument measures the absorbance and calculates and displays UVT.

With respect to installation of any of these systems, consideration of the location of instrumentation is critical for proper operation because WRRFs can periodically have challenges with foam, sediment, or biofilm buildup of algae growth that can interfere with operations of these instruments. Consideration should be given to managing these factors during design and placement of instrumentation that is used in controlling the UV system.

3.4.3 Sleeve Cleaning Methods and Ancillary Facilities

Cleaning UV lamp sleeves can extend their service life and improve system performance and system inactivation efficiency. Ultraviolet reactor manufacturers have developed different approaches for cleaning lamp sleeves. Typical approaches include online mechanical cleaning, which may or may not be coupled with chemical cleaning, and online mechanical–chemical cleaning. Most wastewater systems include automatic wiper systems, which are generally sufficient to maintain lamp sleeve transmittance when coupled with periodic chemical cleaning; chemical cleaning can be done either online or offline, depending on the system selected. When mechanical cleaning systems are provided, the design engineer must include space, power, and, potentially, compressed air to meet the requirements of the specific system, which may use either electric motors or pneumatics to drive the wiper system.

Some UV equipment systems require additional facilities for offline chemical cleaning; for offline chemical cleaning, the reactor is shut down, removed from service, and flushed with a cleaning solution either in a dip tank or cleaning tray. The simplest offline cleaning system is a simple drain tray used to collect drainage from wash-down that can be routed to a sewer or to the head of the facility. Offline chemical systems require water to dilute cleaning chemicals and for safety equipment such as shower/eyewash stations. If a dip tank is required, design should include considerations with respect to design of the lifting system for removing equipment from the channel and placing it into the tank; provisions should be included for draining to a sanitary sewer or back to the head of the plant. Chemical cleaning is typically accomplished using a citric acid or phosphoric acid solution that can be neutralized and disposed after it is spent.

4.0 REFERENCES

American Association of Cost Estimators International (2012) Cost Estimate Classification System—As Applied for the Building and General Construction Industries. Recommended Practice No. 56R-08. TCM Framework: 7.3—Cost Estimating and Budgeting. http://www.aacei.org/non/rps/56r-08.pdf (accessed June 2014).

Institute of Electrical and Electronic Engineers (1993) *Recommended Practices and Requirements for Harmonic Control in Electrical Power Systems*; Standard 519-1992; Institute of Electrical and Electronic Engineers: New York.

U.S. Environmental Protection Agency (2006) *Ultraviolet Disinfection Guidance Manual for the Final Long Term 2 Enhanced Surface Water Treatment Rule*; EPA-815/R-06-007; U.S. Environmental Protection Agency: Washington, D.C.

U.S. Environmental Protection Agency (1969) *Design Criteria for Mechanical, Electric, and Fluid System and Component Reliability*; EPA-430/99-74-001; U.S. Environmental Protection Agency: Washington, D.C.

7

Ultraviolet Project Delivery, Startup, and Commissioning

Gary L. Hunter, P.E., BCEE

1.0	INTRODUCTION	154	5.2	Performance Testing	163
2.0	PROJECT DELIVERY METHODS	154		5.2.1 Testing Protocols	163
	2.1 Design/Bid/Build	155		5.2.2 Testing Duration	164
	2.2 Construction Management at Risk	155		5.2.3 Flow	164
	2.3 Design/Build	156		5.2.4 Water Quality	164
	2.4 Equipment Procurement Methods	156		5.2.5 Analysis of Data	165
3.0	KEY CONSIDERATIONS DURING CONSTRUCTION	158	5.3	Alternative Performance Testing Methods	165
4.0	KEY ACTIVITIES DURING STARTUP	161		5.3.1 Stress Testing	165
	4.1 Commissioning, Startup, and Testing Plan	161		5.3.2 Velocity Profile Testing	165
	4.2 Startup Checks	162	6.0	PERFORMANCE STANDARDS AND SYSTEM ACCEPTANCE	166
	4.3 Functional Acceptance Testing	162		6.1 Power Consumption	166
5.0	KEY ACTIVITIES DURING PERFORMANCE TESTING	163		6.2 Headloss	166
	5.1 Operational Acceptance Testing	163		6.3 Enforcing Manufacturer Guarantees	166
			7.0	REFERENCE	166

1.0 INTRODUCTION

While each owner has unique project delivery and equipment procurement process requirements, this section provides an overview of some of the important decisions when embarking on a UV disinfection project. Specifically, these are choice of the project delivery method and equipment procurement method, which determine how the project will be designed and constructed. Additionally, during the design phase, logistics of construction, startup, and commissioning should be addressed to provide a successful installation for the owner. Contract documents developed by the design engineer for the owner should provide the required information for the contractor to construct the facilities and install the equipment. Depending on the owner, the owner's staff, the design engineer, or a separate organization could serve as the owner's representative during the construction. Depending on the contract, the controls and programing could be provided by an instrument and control (I&C) engineer, a separate systems integrator, or by the owner. After the installation is complete, functional testing of the UV system will need to be conducted followed by system commissioning before the facility is provided to the owner to ensure that it is operationally ready.

2.0 PROJECT DELIVERY METHODS

A summary of significant project delivery and equipment procurement methods that are used for a wastewater UV disinfection system is provided in this section. The three most common methods for delivery of a UV disinfection project are

- Design/bid/build (D/B/B), in which the project is completed in three distinct phases (i.e., design, bid, and construction);
- Construction manager at risk (CMAR), in which a construction manager supervises both design and construction for the owner; and
- Design/build (D/B), in which engineering design and construction are under one contract.

This is not a comprehensive list of delivery methods and there are many variations within these options. An owner faced with choosing a project delivery method should consider several factors in making the decision, including project size, legislative and regulatory requirements, tolerance for risk, schedule, desired level of involvement, and owner resources and capabilities. When these factors are properly evaluated, selection of a project delivery method that best fits the goals and requirements of the owner and the project can be made. Use of a qualified construction manager can aid in developing a project and in making a decision on the project delivery method. The project delivery method also drives the options that are available for equipment selection and procurement (CMAA, 2012). Each method has different equipment procurement strategies, contractual arrangements, and compensation methods; this chapter provides a brief overview of how these strategies align with various delivery methods. There are many resources available to assist in defining the best delivery method and the W/WW Project Delivery Selection Matrix, an online tool hosted by the University of Colorado Boulder at https://dbiarmc.colorado.edu/ is a good reference. The Web site also contains an excellent graphical matrix of comparative information about project delivery and procurement methods.

2.1 Design/Bid/Build

Design/bid/build (D/B/B) is a traditional U.S. project delivery method that customarily involves three sequential project phases: the design phase, which engages an engineer to design the project, including construction drawings and specifications; the bid phase, when a contractor is procured; and a construction phase, when the project is built by the contractor. The D/B/B system remains the most frequently used delivery method for construction projects. Using this method, the owner engages a design engineer to design the project, including construction drawings and specifications. Once completed, the bid package, including the design and bidder's information packet, is presented to interested contractors, who prepare and submit their bids for the work. The owner will select a contractor, typically based on the lowest responsive and responsible bid (for most public work) or some hybrid of price and technical merit. The selected general contractor will then execute contracts with the UV supplier to provide the equipment system and the contractor is responsible for constructing the facility in accordance with the contract documents. The equipment systems may be procured using a variety of methods, which were summarized in Chapter 6. Regardless of the procurement method used for equipment selection and procurement, the design engineer typically maintains some limited oversight of construction and installation of the system and responds to questions about the design on behalf of the owner.

2.2 Construction Management at Risk

Construction management at risk (CMAR) is similar in many ways to D/B/B in that the CMAR acts as a general contractor during construction. That is, the CMAR holds the risk of construction performance and guarantees completion of the project for a negotiated price, which is typically established when the design is somewhere between 50 and 90% developed. In addition to providing the owner with the benefit of preconstruction services, CMAR offers the opportunity to begin construction before completion of the design, which can provide significant schedule advantages. The CMAR can bid and subcontract with the UV equipment provider at any time during the project, often while design of unrelated portions of the project are still not complete. The CMAR and owner often negotiate a guaranteed maximum price based on a partially completed design, which includes the CMAR's estimate of the cost for the remaining design features. Furthermore, CMAR may allow performance specifications or reduced specifications to be used because the process can lead to early agreement on preferred materials, equipment types, and other project features. This process often provides the owner the greatest level of control over UV equipment selection; however, it is still recommended that a life cycle analysis be conducted so that the owner can understand how to budget for operation and maintenance (O&M) of the UV system over the life of the project.

2.3   Design/Build

Design/build (D/B) is a project delivery method that combines engineering design services with construction performance under one contract. Although there are a number of variations on this delivery method, these variations are generally to support different methods of project funding. The D/B project delivery system has grown in popularity because it addresses the limitations of other methods including schedule and point of legal responsibility. With D/B, the owner contracts with a D/B team that performs the complete design of the facility, typically based on a preliminary scope or design presented by the owner. Early in the project, the D/B team will establish a fixed price to complete the design and construction of the facility. Through a prescribed process, D/B offers the opportunity to save time and money because the D/B team is working together from the outset. However, the advantages of the system are offset by a significant loss of control and involvement by the owner and other stakeholders. The primary caution for an owner considering D/B is that the owner should carefully consider the level of involvement it requires with respect to selecting equipment for a successful project, and owners with highly specialized needs may not find it advantageous to turn over responsibility to an outside D/B team. Thus, when an owner is selecting a D/B team for a UV project, he or she should pay particular attention to selecting a team with a successful track record of similar projects.

2.4   Equipment Procurement Methods

The decision on the best equipment procurement method will be based on the available capabilities to manage the procurement and design process; additional considerations are the time sensitivity for project completion and level of risk tolerance. Understanding the project delivery process can provide insight to bid costs that will be provided for equipment systems. For example, a UV equipment supplier may provide a low capital cost for their system in a D/B project, where the objective of the project may be to control construction capital costs; in these cases, the guaranteed replacement parts costs for lamps may be substantially higher that in a traditional D/B/BJ project. Some of these issues can be overcome by requiring an evaluated UV equipment package bid that also accounts for life cycle costs of the project.

Because UV disinfection equipment systems are substantially different among manufacturers, there are a number of equipment selection/procurement procedures that have been used to compare different systems and meet the procurement requirements of specific owners. In the United States, most owners purchase or select the UV equipment before the UV facility design is complete. Prepurchase or preselection of the UV equipment enables the design engineer and the UV manufacturer to coordinate during the detailed, final design phase to consider manufacturer-specific design recommendations. Sometimes the equipment is pre-selected and the UV equipment manufacturer is included in the construction contract. Other procurement methods (e.g., base-bid and contractor selection of equipment) are also used, but these methods are less common. It is critical to meet with the owner's purchasing department to ascertain their policies and procedures because they will determine the possible project delivery methods and procurement options for a particular project.

Where an owner has requirements for competitive bidding, the equipment procurement process may require preparation of multiple designs to allow bidding of multiple UV equipment systems. In this case, the final equipment selection is often left up to the general contractor, with the decision typically based on lowest capital cost. Whereas this method ensures a competitive capital price for the owner, it does not always take into account differences in long-term (O&M) costs, unique UV equipment features, or support services provided by the competing UV manufacturers. There are other equipment procurement options that typically include the following: (1) sole source of equipment with design around the selected equipment; (2) design around a single supplier, allowing others to bid as long as they prepare a revised design; or (3) competitive preselection of equipment with the design being performed around the selected equipment. In a sole-source procurement scenario, the design can be based on a single UV manufacturer and the equipment is procured by the general contractor from the single named supplier in the bid documents. This simplifies design, but often results in the owner paying a premium equipment price. To obtain the best price using this method, it may be possible to prenegotiate a scope of supply and equipment price with the selected UV manufacturer in advance of the project bid. This method is not always allowed for in many publicly funded projects, which typically require significant equipment to be competitively bid. Alternatively, the design could be developed around a single supplier, allowing other suppliers to bid as long as they prepare a revised design, if selected.

Prequalification or preselection of equipment can help streamline the process, but it also calls for a comprehensive assessment at the early stage of design. In this case, a desktop assessment for UV technology selection is conducted in the preliminary engineering report to preselect equipment and the selected system is written into the general contractor's contract. An interim agreement is required between the owner and the UV manufacturer to fix the bid price for a prescribed time period so that the agreement and bid price can be transferred to the general contractor selected for the project. Upon assignment, the contractor assumes all responsibilities for performance of the UV manufacturer under terms of the interim agreement between the owner and the manufacturer. There may be restrictions on this process for some owners; thus, an alternative to this process is use of an evaluated bid as part of the general contractor's contract so that the UV equipment can be selected based on lowest life cycle cost.

Ultraviolet equipment can also be prepurchased and this procurement method allows for a formal evaluation of competing UV equipment manufacturers through a request for proposal (RFP) and selection process in advance of the overall project bid. Whereas this equipment procurement method may allow the owner to avoid the sales tax that would otherwise be paid by the contractor, it does burden the owner with additional project responsibilities and risks with respect to coordinating the UV manufacturer's scope of supply to avoid any effects on the overall construction project, including schedule delay claims by the general contractor.

There are a variety of methods that can be used within the framework of the various project delivery methods and a summary of the UV equipment procurement methods along with generalized advantages and disadvantages is provided in Table 7.1; it is important to note that there may be variability in some of these factors based on the regulating authority for public projects. This summary of procurement methods is not intended to be a comprehensive list of all possible scenarios, and other terminology may be associated with the named methods.

3.0 KEY CONSIDERATIONS DURING CONSTRUCTION

This section describes many of the activities that will need to be completed during construction, startup, and commissioning. Figure 7.1 provides an overview of activities that occur during startup and commissioning phases

TABLE 7.1 Summary of equipment procurement methods.

Procurement method	Advantages	Disadvantages
Base-bid (based on preselected system)	Single design around selected UV system UV system price can be prenegotiated with UV supplier Lower engineering design cost Contractor handles all pricing and coordination with UV supplier	Owner may pay a premium price for preselected vendor Municipal procurement regulations may not allow this option Difficult to prevent supplier from "packaging" equipment
Contractor selection	Allowed by most procurement codes UV equipment disputes are responsibility of contractor Competitive bids received based on capital cost only	Contractor selects based on low capital cost, not lowest life cycle cost or technical criteria May require multiple UV designs to account for design differences for competing UV technologies
Owner prepurchase	Competitive bids based on capital cost, O&M cost, and/or technical criteria Single design around selected UV equipment UV equipment pricing is defined at start of construction No municipal sales tax Shorter project schedule possible if equipment fabrication occurs during design and bidding phases	Requires two bid documents—equipment procurement RFP and general contractor construction documents Owner must coordinate UV equipment contract; equipment disputes resolved by owner Municipal procurement regulations may not allow this option General conractor is not a single point of responsibility for equipment

of implementation of a wastewater UV disinfection project. As shown in Figure 7.1, the activities need to be coordinated between the design engineer, owner's representative, contractor, controls programmer, and UV manufacturer. Each UV manufacturer has a standard set of activities that they provide during construction, functional testing, and commissioning that are specific to the type of UV system.

Construction includes the mechanical, civil, structural, electrical, and instrumentation activities of creating the project based on issued plans and specifications. Before and during the process, the following activities should occur:

- Submittals are reviewed and approved;
- Requests for information are answered;

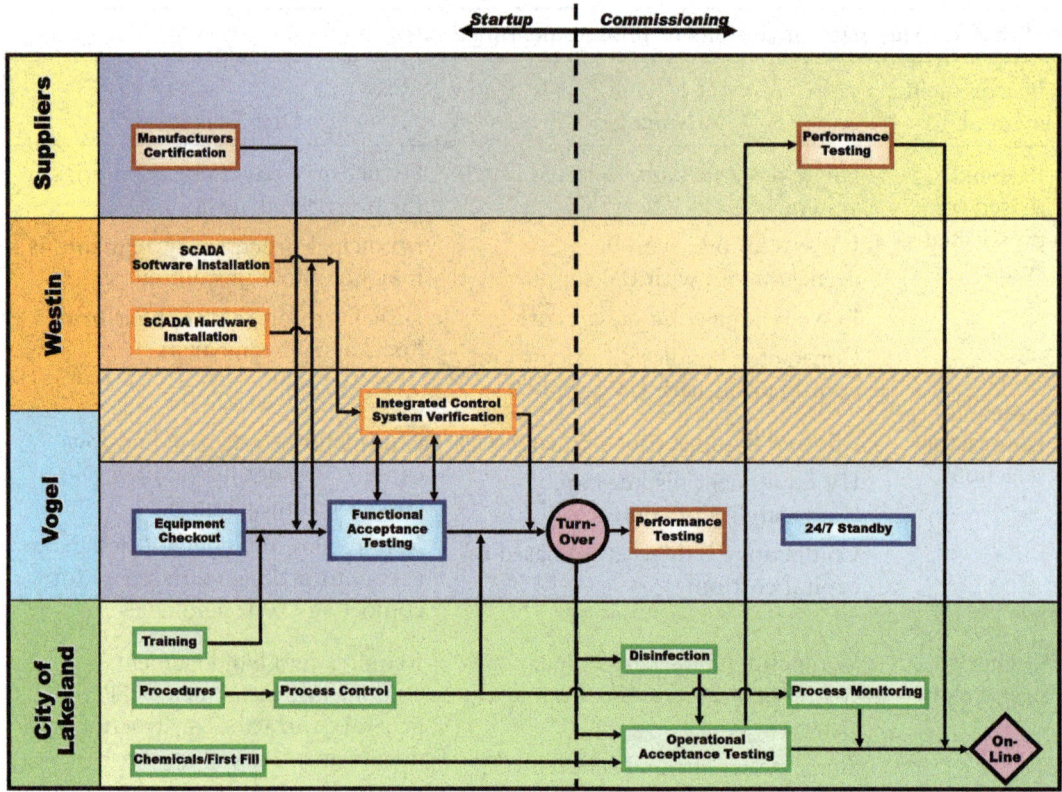

FIGURE 7.1 Startup and commissioning activities.

- Equipment and control systems are factory-demonstration tested;
- Leakage testing is performed on piping, tanks, and structures/reactors that will hold liquid;
- Pre-operational tests are conducted on mechanical equipment (heating, ventilating, and air-conditioning; valves; actuators; pumps; and blowers); piping systems are pressure-tested and pipes cleaned/flushed; instrumentation equipment includes instrument calibration, loops, programmable logic control (PLC) and power control center controls and logic, public address system, closed-circuit television system, fire alarm system, fiber optic cables, and electrical equipment (National Electrical Testing Association), conductors, and motors);
- Equipment manufacturers' installation certifications;
- Commissioning plan received and approved;
- Design engineer draft and final operations manual is received and approved; and

- Record computer-aided design drawings, master equipment list, and maintenance managed system items.

It is recommended that the reader consult with the Association of General Contractors for additional information on the construction process. During construction, factory acceptance testing (FAT) needs to take place. It is imperative that FAT occurs for most UV systems so the I&C engineer can confirm that PLC programming meets the specifications. The I&C engineer needs to visit the factory after the manufacturer indicates that the PLC programming is complete and before the PLC is shipped. The I&C engineer then witnesses a demonstration of each PLC screen and program activity. Notes should be made regarding deficiencies so the manufacturer can correct them before shipment.

4.0 KEY ACTIVITIES DURING STARTUP

During the startup phase of any UV disinfection project, a number of activities need to occur including preparation of a commission startup and test plan, conducting startup checks, and functional testing.

4.1 Commissioning, Startup, and Testing Plan

As part of the construction activities, a commissioning, startup, and testing plan needs to be developed. This plan should discuss all the startup activities that will be conducted by the manufacturer and contractor and when these activities are to occur. This plan needs to be reviewed by the owner and the owner's representative to ensure that the staging of the activities will occur at an appropriate time for the facility. For instance, the plan should have provisions regarding postponing startup during peak flow and solids loading events. A number of activities (while not inclusive) are suggested that may be included in the startup plan. These are as follows:

- Narrative testing and startup plan—including all contractual and regulatory requirements
- Summary of objectives for startup checks, functional testing, functional acceptance testing, commissioning, and performance testing
- List of items and systems that will undergo startup with reference to contract documents
- Electrical and instrumentation inspections
- Schedule for various testing activities

- Safety and emergency response requirements
- Organization chart and responsibilities for each party
- Startup and testing record plan
- Sampling and reporting plan
- Plan for reuse and disposal of wastewater used or generated during the testing
- Description of any temporary facilities required
- List of any chemical and who will be providing

After the UV manufacturer certifies that the equipment has been installed based on their specifications, then a series of startup and functional testing checks are conducted based on the requirements contained in the construction documents.

4.2 Startup Checks

Before water is introduced to the UV system, a number of startup checks need to be completed. The equipment/channels should be inspected and ensured that no foreign matter is in the channel or reactor. In addition, comprehensive input/output checks need to occur during the startup activities. These checks are necessary to verify integrity and accuracy of signals between UV reactors; appurtenances such as gates, valves, and analyzers; local control panels; the master control panel; and any associated supervisory control and data acquisition platform. Additionally, a check will need to be made of the status of each lamp and it should be repaired and replace as appropriate.

4.3 Functional Acceptance Testing

Functional acceptance testing should include activities that run the entire UV system through its "paces". This means ramping the flow up and down, which should trigger the activation of additional banks and/or treatment trains, with associated closing/opening of gates and valves. Depending on actual flow variation as experienced by the utility, this may require the use of dummy signals to the control panels. Monitoring of the calculated dose, status of lamps, status of banks, power consumption, UV intensity, flow, and UV transmittance (UVT) during this testing is important. It is particularly important to check that the calculated dose equation is correct and yields the correct dose under variable conditions as experienced during testing. The testing may also reveal previously unforeseen scenarios where transient phenomena lead to unacceptable performance (e.g., a gate opening more slowly than the flow is ramping up, and causing flow to exceed its validated range within active

UV trains). It is not unusual for PLC programming to be modified to suit the specific conditions of the facility as a consequence of the testing.

5.0 KEY ACTIVITIES DURING PERFORMANCE TESTING

All the activities described in Section 4.0 should be completed before the activities in this section are started. This section summarizes activities that need to occur on-site (at the location of the UV system) and is often referred to as the "activation" part of the project, and includes such activities as

- Contractor's punch list completed and field-accepted;
- Regulatory notice of field acceptance;
- Commencement of warranty period; and
- Process optimization, operational acceptance, and process performance test.

Key items that affect the UV system are the conduction of operational and performance testing requirements, which are discussed in the following subsections. Requirements for this testing would have previously been incorporated to the contract documents.

5.1 Operational Acceptance Testing

During operation acceptance testing, the UV system will run in all operational modes (manual, remote-manual, and automatic) based on the requirements of the contract documents. Operation requirements during the test should be identified within the contract documents. Duty and standby equipment should be alternated during the testing so that all the equipment is shown to be meeting the requirements of the system.

5.2 Performance Testing

The overall goal of performance testing is to show that the installed system meets or exceeds the performance requirements outlined within the construction documents. The contract specifications should describe the testing requirements, protocols, and how data will be summarized and compared to the design requirements.

5.2.1 Testing Protocols

Before performance testing is conducted, testing protocols need to be developed to ensure that the testing that is completed meets the requirements of the contract documents. Each installation may have specific requirements that are established for that location. The specific goals of the testing that are identified within the contract documents should also be included in the testing protocols. Goals may also include requirements related to headloss and power supply.

The protocols should address key issues like power interruptions, water quality that does not meet specified values, flow in excess of the contract documents (both in amount and variability), data that will be collected during the test (electrical power draw, UV system controls, bacterial sampling, UVT, particle size and total suspended solids, and other items). The protocols should provide a summary of how the data will be analyzed to show conformance with the contract document requirements. The protocols should also outline what data should be collected and various analytical procedures for the microbiological samples that are collected as part of the test.

5.2.2 Testing Duration

Length of the testing will typically be established by the contract documents and will be determined by the design engineer. This may range from several hours to several days depending on the exact performance requirements. For instance, if the goal of the testing is to examine performance at peak flow conditions, then a shorter test period may be desirable because of availability of flow. If the testing is to evaluate operation of the control system, a longer test may be warranted (exceeding 30 days).

5.2.3 Flow

Performance guarantees are generally established at the peak design peak flow, which may not be realized until sometime well into the future. As such, the testing protocol (or even process design in terms of the number of channels or trains) should be established to determine how the performance test will be run when there are lower than expected flows. One approach is to have the design engineer specify flowrate based on either gallons per minute per lamp or liters per minute per lamp. This approach allows for easy adjustments to be made in the field.

5.2.4 Water Quality

The water quality at average and peak conditions is typically defined in the design documents. This may include requirements for a variety of parameters including UVT, effluent solids, influent UV channel bacteria, effluent UV channel bacteria, facility effluent biochemical oxygen demand/chemical oxygen demand and other water quality requirements. Contact documents will also establish the frequency at which these parameters will be collected during the testing program.

One of the considerations during the test may include adjustment of water quality to simulate the peak operational requirements established as part of the contract documents. While this may be conducted using a number of methods, consideration should be given to where the disinfected water is being discharged (and whether there is a potential for a permit violation because of this adjustment), if the treated water can be discharged back to the influent of the facilities, and whether the facilities have the capacity to handle the flow surges that may occur.

5.2.5 Analysis of Data

Depending on the time duration and protocols listed in the contact documents, the analysis of the data may vary. For short test periods, the analysis may only examine the peak requirements, whereas, if the testing duration extends over a 30-day period, a geometric mean of the effluent data may need to be calculated. Therefore, the contract documents will establish how the data will be analyzed to show compliance with the requirements.

5.3 Alternative Performance Testing Methods

The design engineer may provide for alternative approaches to performance testing in the contract documents. These approaches may be considered when there is no physical way of collecting samples during the testing to verify performance. Therefore, it is anticipated that these approaches would be used infrequently.

5.3.1 Stress Testing

A number of alternative performance testing approaches have been examined to reduce the overall cost of the performance test. One such approach is to conduct a stress test. A stress test would be accomplished adding a surrogate such as MS2 or T1 instead of adjusting the operations of the treatment facility because that may result in a compliance issue. Stress testing may also be examined during wet weather or peak flow events if it is included within the specification.

5.3.2 Velocity Profile Testing

One method of assessing reactor performance is to examine the velocity profile within a specific reactor configuration. While it is a viable tool, it may not always be a good predictor of UV system performance; therefore, careful examination of the data needs to be conducted. Velocity testing has mostly been replaced with spot check bioassay because it is a more scientific method of assessing overall performance.

6.0 PERFORMANCE STANDARDS AND SYSTEM ACCEPTANCE

The contract documents should establish the various requirements for all the elements associated with the performance testing. This would include any additional power monitoring that may be required. These documents will also identify the key parameters and penalties that may be imposed for the failure of a test. Once all of the performance standards have been met, then the system can be turned over to the owner for operation.

6.1 Power Consumption

It is interesting to note that to check actual power consumption, a power meter can be attached to the UV system and power information collected at various power settings. This is somewhat independent of performance test results.

6.2 Headloss

One of the key parameters to verify during the UV system performance test is headloss. If headloss results are not correct, then the system will not be able to treat the design flowrate. Care must be taken in collecting headloss data because minor variations in measurements can lead to larger variations when the data are analyzed.

6.3 Enforcing Manufacturer Guarantees

The contract documents for the UV disinfection system establish the performance requirements and the overall penalties for noncompliance. The overall challenge is to collect data during the performance test that is accurate and representative. Therefore, modifications may need to be made at the time of the test to account for flow and water quality.

7.0 REFERENCE

Construction Management Association of America (2012) An Owner's Guide to Project Delivery Methods. https://cmaanet.org/files/Owners%20Guide%20to%20Project%20Delivery%20Methods%20Final.pdf (accessed June 2014).

8

Operational Considerations

Jay Swift, P.E., and Chad Newton

1.0	ROUTINE OPERATION	168	5.3	Control Cabinets/Programmable Logic Controllers	178
2.0	SAFETY	168	5.4	Supervisory Control and Data Acquisition System Integration	178
	2.1 Electrical Hazards	169			
	2.2 Ultraviolet Radiation Hazards	169	6.0	MAINTENANCE CONSIDERATIONS	180
	2.3 Lifting Hazards	170		6.1 Lamp Replacement	180
	2.4 Chemical Hazards	170		6.2 Sleeve Replacement	182
	2.5 Other Hazards	171		6.3 Ballast Replacement	182
3.0	PROCESS MONITORING	171		6.4 Sleeve Cleaning	182
	3.1 Ultraviolet-Intensity Sensors	171		6.5 Channel Cleaning	183
	3.2 Ultraviolet Transmittance Meters	173	7.0	ANCILLARY SYSTEMS	184
	3.3 Flow Monitoring	173		7.1 Chemical Cleaning Tanks/Stations	184
	3.4 Level Monitoring	174		7.2 Air Compressors/Blowers	185
	3.5 Temperature Monitoring	175		7.3 Bank/Module Lifting	185
	3.6 Warnings and Alarms	175	8.0	TROUBLESHOOTING	185
4.0	OPERATIONAL STRATEGIES	175		8.1 Electrical Issues	185
	4.1 Energy Conservation	175		8.2 Low Ultraviolet Transmittance	185
	4.2 System Redundancy (Multiple Banks and Channels)	177		8.3 Hydraulics	187
				8.4 Upstream Process Effects	187
5.0	ELECTRICAL AND CONTROL SYSTEMS	177	9.0	REFERENCES	188
	5.1 Power Requirements and Power Redundancy	177			

1.0 ROUTINE OPERATION

Generally, a UV disinfection system should be operated in accordance with the operation and maintenance (O&M) manual provided by the manufacturer and/or engineering designer. Depending on the facility, the system may be operated in flow-paced, dose-paced, or manual mode, as described in more detail in this chapter. To ensure that the UV system functions properly, a number of routine maintenance activities, as described in this chapter, should be implemented.

Typical operations tasks are summarized in Table 8.1. During normal operation, secondary or tertiary effluent is passed through the UV reactor(s) at flowrates and transmittances within the design range of the system. The system will typically be operated with monitoring systems in use to allow alarms to indicate problems with the system.

2.0 SAFETY

Safety is one of the most important items to consider in O&M of UV disinfection systems. Safety equipment provided by the manufacturer will need to be used (as described by the manufacturer) to ensure that injury does not occur. Additional information is provided in *Safety, Health, and Security in Wastewater Systems* (WEF, 2012); however, a summary of the key health

TABLE 8.1 Typical operational tasks (WEF, 2006).

Frequency	Recommended tasks
Daily	• Perform overall visual inspection of UV reactors. • Ensure system control is on automatic mode (if applicable). • Check control panel display for status of system components and alarm status and history. • Ensure that all on-line analyzers, flow meters, and data-recording equipment are operating within parameters. • Review effluent data.
Weekly	• Initiate manual operation of wipers (if provided).
Monthly	• Check lamp run time values. Consider changing lamps if operating hours exceed guaranteed life or UV intensity is low.
Semiannually	• Check ballast cooling system for leaks or unusual noise from cooling fans. • Check operation of automatic and manual valves.

and safety considerations in operating the UV systems are summarized in the following subsections.

2.1 Electrical Hazards

To prevent electrical hazards, all safety and operational precautions required by the current National Fire Protection Agency National Electric Code, Occupational Safety and Health Administration, local electric codes, and the UV manufacturer should be followed and include the following precautions (U.S. EPA, 2006):

- Proper grounding,
- Lockout/tagout procedures,
- Use of proper electrical insulators, and
- Installation of safety cutoff switches.

According to the *Ultraviolet Disinfection Guidance Manual for the Final Long Term 2 Enhanced Surface Water Treatment Rule* (U.S. EPA, 2006), proper grounding and insulation of electrical components are critical for protecting operators from electrical shock and protecting the equipment. To minimize electrical hazards, ground-fault-interruption (GFI) circuitry should be provided with each module. For a GFI to function properly, the UV reactor ballast must not be electrically isolated from the main power supply. If it is isolated with a transformer, then the GFI must monitor the secondary side of the transformer for any current leakage.

Provisions enabling the UV reactors to be isolated and locked out for maintenance, both hydraulically and electrically, should be included in the design. Control of all lockout systems should remain local.

2.2 Ultraviolet Radiation Hazards

Ultraviolet radiation can cause several safety issues like burns and damage to eyes and skin. To protect staff from damage associated with exposure to UV light, the following precautious should be implemented:

- While channel covers/grating are typically designed to prevent or mitigate potential UV light exposure during operation of the system, warning signs regarding UV radiation should be posted;
- Lamps should not be operated in air to prevent overexposure of skin and eyes to UV radiation. This means that UV systems should be equipped with safety interlocks that will automatically shut down lamp modules if they are taken out of the reactor or the water level

falls below critical levels (typically, the top of the lamps in the reactor for horizontal systems) (WERF, 2007);

- Regular maintenance on the system must be performed with UV lamps off, if possible; and
- If maintenance needs to be performed while the system is operating, proper eye/face shields and skin protection must be used; eyes should never be exposed to UV radiation and it is important to not look directly into a UV lamp that is on. (It should be noted that, in closed-vessel UV systems, there is minimal exposure to UV light compared to open-channel UV systems, although caution must still be exercised if sensors are removed for inspection from sensor ports.)

2.3 Lifting Hazards

Ultraviolet modules and banks can be heavy and may require cranes or davits for lifting for maintenance. While the design of lifting systems that are required for some UV systems should follow all local, state, and federal (e.g., National Electrical Code) guidelines, operators should also be trained on using manufacturer recommendations on removal of UV equipment. In all instances, caution should be exercised to avoid risk of injury when lifting and moving modules and banks and cranes should be operated by experienced staff who have received all necessary training for safely operating the cranes.

2.4 Chemical Hazards

Ultraviolet lamps contain mercury that can be liberated if the lamps are accidentally broken. Mercury disposal may fall under state and federal waste disposal rules; therefore, it is important to consult both manufacturer guidelines and local and state requirements regarding proper handling procedures and proper disposal for used or defective lamps. Additional considerations for handling mercury include provisions for sufficient air exchange and/or ventilations in working rooms. Contact with mercury must be avoided, and released mercury must be contained with a mercury containment kit. In some cases, disposal of the released mercury can be done at the same time as UV lamp disposal. Operators must refer to the material safety data sheet (MSDS) provided by the UV supplier for safety guidelines.

Cleaning solutions typically contain acids and should be handled with appropriate precautions. Cleaning solutions may contain phosphoric or other moderate (citric) to strong (hydrochloric) acids; as such, gloves and eye protection are necessary (consult the relevant MSDSs for specific recommendations on chemical handling and disposal).

2.5 Other Hazards

Other hazards include the following:

- Ultraviolet lamps remain hot for a considerable time after they have been turned off. Lamps should be allowed to cool off after use for at least 5 minutes for low-pressure lamps and 15 minutes for medium-pressure lamps;
- There is a risk of flooding during peak flow events in open-channel systems. Protection with interlocks to shut off electricity based on water level may be needed to protect equipment from water damage (short-circuiting, damaging equipment, and creating an electrocution hazard) if equipment is energized and is flooded;
- Tripping and slipping hazards may exist around UV channels. This is especially a risk when grates are removed; and
- Staff should take care in handing broken lamps because broken glass can pose a risk of staff being cut.

3.0 PROCESS MONITORING

Ultraviolet disinfection control systems use process instrumentation for monitoring, reporting, and control. This section describes the most common process instrumentation associated with UV disinfection systems, potential uses for process control, and O&M factors. The section also provides guidance for disinfection system warnings and alarms.

3.1 Ultraviolet-Intensity Sensors

Ultraviolet-intensity sensors provide information related to UV intensity at various locations in the reactor. Typically, UV manufacturers provide one intensity sensor per reactor or bank and determine the location of the sensor based on the intended UV control system. The UV-intensity measurement at a specific point reflects changes in lamp output related to lamp power setting, lamp aging, quartz sleeve fouling and aging, and changes in UV transmittance (UVT) (WERF, 2007). A single sensor can only report the UV intensity at a single location. It cannot report an average intensity within the entire reactor or bank unless strategically placed by the manufacturer. Because of the non-uniform UV-intensity profile within a UV disinfection system, it is problematic to estimate the average UV intensity or dose from a single intensity sensor in the reactor. Ultraviolet manufacturers typically conduct bioassays to correlate the measured UV intensity at the selected sensor location to average reactor UV intensity.

Sensor locations closer to UV lamps will disproportionately respond to changes in UV lamp output and quartz sleeve fouling. Sensor locations further from lamps will disproportionately respond to changes in the UVT of the wastewater. Intensity sensors must be precisely placed if the intent is to incorporate changes to wastewater UVT in the intensity measurement. If UVT is monitored separately with a transmittance monitor, the intensity sensor may be placed more closely to UV lamps and sensor placement is not as critical.

Some UV system manufacturers have developed dose algorithms for system-display, reporting, and dose-pacing purposes, which have been calibrated to intensity sensor readings, with or without the inclusion of UVT sensor data. Other UV control systems use intensity sensors to report relative differences in performance or for alarming only. Intensity sensors can be used to estimate lamp aging or sleeve fouling effects over time. A sensor reading is recorded under new lamp and unfouled conditions, with subsequent recording of sensor readings over time. Sensor readings can be compared pre- and post-cleaning to estimate cleaning effectiveness. They can be compared over time on clean systems to discern the need to replace lamps and/or sleeves. Low-intensity warning/alarm setpoints can be used to warn against potentially inadequate disinfection.

Intensity sensors require frequent cleaning and calibration verification to provide accurate results. Intensity sensors require periodic cleaning to avoid scaling or fouling on the UV-intensity receiving surface, and the cleaning frequency will depend on the site-specific water quality. Calibration verification may consist of maintaining a single unused sensor as a reference standard and inserting that sensor in place of operational sensors to measure drift. When drift of operational sensors become excessive, a calibration factor can be applied to the control system or those operational sensors can be recalibrated or replaced. Some manufacturers provide intensity sensors that may be field calibrated (typically when intensity readings are not used for dose control or verification), whereas many sensors may only be recalibrated at the factory. For reclaimed water systems, the National Water Resources Institute (NWRI) requires that intensity sensors must be calibrated at least monthly (NWRI and WRF, 2012). For drinking water UV disinfection systems, the U.S. Environmental Protection Agency (U.S. EPA) recommends that intensity sensor readings be verified against the reference sensor monthly. The *Ultraviolet Disinfection Guidance Manual for the Final Long Term 2 Enhanced Surface Water Treatment Rule* (U.S. EPA, 2006) provides detailed calibration procedures for intensity monitors.

Although in practice, some facilities have found the use of intensity sensors in wastewater problematic, as a result of design issues, poor accuracy, fouling, lack of robustness, cleaning efficacy and sensor reading drift challenges, UV intensity readings together with rigorous calibration and maintenance along the lines recommended by NWRI or the U.S. EPA will result in greater confidence in the UV system performance. The rigor of intensity sensor calibration and maintenance in low-dose wastewater treatment applications will depend on the control system and maintenance recommendations provided by the UV system manufacturer, as NWRI or U.S. EPA standards are not typically required by regulators. For other UV control systems that do not consider intensity readings, the sensor calibration and maintenance schedule may be determined by manufacturer's recommendations or selected by facility staff.

3.2 Ultraviolet Transmittance Meters

Ultraviolet transmittance is one of the most critical water quality parameters in determining disinfection effectiveness. Industrial discharges and wet-weather events often significantly affect UVT. It is important to note that changes to UVT cannot be discerned by the human eye. Just because water exhibits color (e.g., a greenish tinge resulting from algae) does not indicate that the water also will exhibit a low UVT. Conversely, certain chemicals are colorless, yet have high UV-absorbing characteristics, which result in significant effects on UVT. Simply because an effluent is crystal clear does not indicate high UVT.

Ultraviolet disinfection systems may be equipped with online UVT sensors. During operation, the disinfection control system may regulate dose delivery in response to UVT in an effort to minimize energy requirements. Another option is to maintain the system dose required for the minimum design UVT and use online transmittance monitoring to trigger alarms.

Ultraviolet transmittance sensors must be maintained and calibrated according to manufacturer recommendations. The NWRI and WRF guidelines (NWRI and WRF, 2012) require that water reclamation UV disinfection systems verify the accuracy of online transmittance sensors weekly by comparison with a calibrated laboratory UVT analyzer. The frequency of transmittance sensor calibration and maintenance in low-dose wastewater treatment applications will depend on whether the control system provided by the UV system manufacturer relies on UVT sensors for dose pacing. At a minimum, the cleaning, maintenance, and calibration recommendations of the sensor manufacturer should be followed.

3.3 Flow Monitoring

Ultraviolet disinfection control systems typically use flowrate information to match the quantity of UV equipment in service, or lamp power, with the current flowrate. However, flow instrumentation is not typically provided by UV disinfection system manufacturers. Data from a facility effluent flow meter is provided as an input to the disinfection control system. The data may be relayed through the facility supervisory control and data acquisition (SCADA) system or, particularly at smaller facilities, there may be a direct data connection between the effluent flow meter and the disinfection control system. Magnetic flow meters and Parshall flumes with level instrumentation are the most common effluent flow meter types.

Facility operators should consider how the recorded effluent flowrate corresponds to the flow through the UV disinfection system. These flowrates may differ because of the addition or diversion of various recycle or service water streams. If the difference in flowrate is significant or highly variable, a more accurate UV disinfection flowrate could be calculated in a programmable logic controller (PLC) by adding or subtracting other measured or estimated flowrates from the recorded effluent flow, with the calculation result sent by the facility SCADA system to the disinfection control system. Operators should refer to the flow meter manufacturer's manual for maintenance and calibration requirements.

3.4 Level Monitoring

In open-channel UV disinfection systems, the wastewater within the channel must be maintained at a nearly constant level with little allowable fluctuation. The wastewater level is typically controlled by a mechanical counterbalanced gate or by fixed or adjustable weirs downstream of the UV reactors. Level control in open-channel UV systems is required to prevent

- Too high a water level resulting in an undisinfected layer of water passing above the lamps or flooding of the channels or nonwetted disinfection system components, such as ballasts; and
- Too low a water level resulting in UV lamps operating in the air instead of submerged, which may damage the lamps through overheating or expose staff to UV radiation.

Ultraviolet disinfection systems are typically designed to ensure that the water level will remain within the required range. However, open channels may be equipped with level monitoring instrumentation to produce alarms for protection against unexpected conditions.

Discrete (e.g., float switches) or analog (ultrasonic level sensors, pressure transmitters, radar level sensors, etc.) level instrumentation may be used to monitor level in open UV channels. Low- and high-level setpoints in the disinfection control system would trigger alarms and/or equipment shutdown. Closed-vessel UV disinfection systems must remain full of water at all times and, therefore, the connecting piping systems are typically designed to ensure full flow. Operators should refer to the level instrumentation manufacturer's manual for maintenance and calibration requirements.

3.5 Temperature Monitoring

Most UV disinfection lamps rely on flowing wastewater for lamp cooling because of the high operating temperatures of the lamps. Medium-pressure and low-pressure high-output UV lamps operate at the highest temperatures. Ultraviolet disinfection systems are typically designed to ensure that the expected range of design flowrates will be sufficient for lamp cooling. However, temperature sensors may be provided by UV system manufacturers to prevent overheating. Flowrates in pumped systems could drop to zero if the pumps are not operating.

Discrete (e.g., temperature switches) or analog (temperature sensors) instrumentation may be used to monitor temperature in UV reactors. High-temperature setpoints in the disinfection control system would trigger alarms and/or equipment shutdown. Operators should refer to the temperature instrumentation manufacturer's manual for maintenance and calibration requirements.

3.6 Warnings and Alarms

Control systems provided by UV disinfection system manufacturers typically provide a variety of warnings, minor alarms, and significant and critical alarms. The terminology and categorization will vary between manufacturers. The warnings and alarms may be locally annunciated at the UV system control panel(s) or they be transmitted to the facility SCADA system for remote annunciation and historical logging. For facilities that are not staffed around the clock, it is recommended that, at a minimum, significant/critical alarms are connected to a remote annunciation system (such as an auto-dialer) to notify off-site staff of the alarm. Some systems may be configured so that UV control system will automatically shut down UV equipment upon certain critical alarms. Table 8.2 provides a list of typical UV system warnings and alarms

4.0 OPERATIONAL STRATEGIES

4.1 Energy Conservation

Ultraviolet disinfection is a significant energy consumer at water resource recovery facilities (WRRFs), so strategies to conserve energy are often central

TABLE 8.2 Typical alarm conditions for UV disinfection systems (adapted from U.S. EPA [2006]).

Alarm name	Alarm category	Description
Lamp age	Minor alarm	Lamp end-of-life per defined operational lamp life
UV sensor calibration	Minor alarm	UV sensor calibration interval reached
Lamp/ballast failure	Minor alarm	Failure of a single lamp or ballast
Multiple lamp/ballast failures	Significant alarm	Failure of adjacent lamps/ballast or failure of >5% of lamps/ballast in a single reactor
Low UV intensity	Significant alarm	Measured UV intensity is below design/validated value
Low UV transmittance	Significant alarm	Measured UVT is below the minimum design value
Low UV dose	Significant alarm	The calculated UV dose is below the design/validated dose
High flow rate	Significant alarm	Measured effluent flowrate exceeds the UV design capacity
High liquid level	Significant alarm	There may be an undisinfected layer of water above the lamps, or potential flooding
Mechanical cleaning system failure	Significant alarm	Failure of the lamp cleaning system
Low liquid level	Critical alarm	Potential for exposed lamps or overheating
High temperature	Critical alarm	Overheating occurring

to operation. A key element of energy conservation is through the use of flow and dose pacing. Flow and dose pacing typically are incorporated to a UV system by the manufacturer. Flow pacing is modulating the power or number of lamps in response to wastewater flow. Dose pacing, typically considered the most energy-efficient means of operational control, is modulating the power to the UV lamps to achieve the desired dose based on three to four factors: flow, lamp power, UVT, and, in some cases, UV intensity. However, excessive lamp cycling may reduce lamp life; thus, many operators will leave lamps running for a minimum duration each time a lamp is turned on.

The controller often will be designed to cycle operational banks and channels in an attempt to age all of the lamps at a similar rate. The designer should check the programmed algorithm and verify that the system is operating within the validated range. If the disinfection permit is seasonal, then significant energy can be saved by shutting down the UV system during the non-disinfection season.

4.2 System Redundancy (Multiple Banks and Channels)

A key aspect of ultraviolet disinfection reactor operation is system redundancy, which has been addressed in Chapter 5. If redundant or standby reactors are provided as part of the system design, it is important for operations staff to exercise this equipment so that it is ready to be used if and when it is needed.

5.0 ELECTRICAL AND CONTROL SYSTEMS

In its essence, UV disinfection is a technology that converts electrical power into UV light energy to inactivate microorganisms in the wastewater. High-quality and reliable electric power is essential for successful disinfection. Furthermore, many UV disinfection systems are provided with control systems that are needed to monitor and adjust parameters related to dose and warn facility operators of inadequate disinfection.

Some WRRFs have installed power meters at the motor control center to independently record incoming power quality and total power consumption of the UV disinfection system. Total power consumption data can be useful in trending UV system performance and troubleshooting lamp ballast failure or excessive power use. Power quality data, such as historical records of voltages on each phase for three-phase power, may be needed for identifying and troubleshooting power quality issues with the electric utility.

To prevent electrical hazard, all safety and operational precautions required by the electric code should be followed.

5.1 Power Requirements and Power Redundancy

Both power quality and redundancy are important design considerations for any UV system to provide continuous disinfection. Discussion of the design considerations for power quality and redundancy are provided in Chapter 6. However, it is important to note that any electrical systems should be inspected and maintained according to the manufacturer's recommendations. It is also important to implement an operational program that includes periodic startup or testing of generators or other electrical equipment that is part of the electrical backup system.

5.3 Control Cabinets/Programmable Logic Controllers

Programmable logic controllers, or customized control boards, are typically incorporated to UV system control panels provided by UV system manufacturers. Control boards/PLCs allow control systems to monitor the status of each lamp, module, or reactor, the UV intensity in each reactor or module, and operating hours on each lamp or module, and provide warnings or alarms to operators about equipment failure or end-of-life. The control boards or PLCs may also contain algorithms and equations for flow pacing and dose pacing by controlling lamp output.

The power services to UV reactor components may differ. For example, the control cabinets may be provided with three-phase, 480-v AC service and include stepdown transformers to provide single-phase, 240-v AC for lamps or 120-v AC service for instrumentation, such as an online UVT analyzer.

Control panels and cabinets provided by UV system manufacturers may require heating or dehumidification to avoid condensation or air-conditioning to avoid overheating, particularly if lamp ballasts are mounted in the cabinets or the cabinets are exposed to direct sunlight.

5.4 Supervisory Control and Data Acquisition System Integration

Ultraviolet disinfection systems typically communicate with facility SCADA systems, particularly at larger installations. The communication will include UV control cabinets sending system status and alarm information to the SCADA system. Field instrumentation data from outside the scope of the UV system, such as effluent flow meters or UVT analyzers, can be sent to the UV control cabinets via the facility SCADA system. The SCADA system can also be programmed to provide UV control cabinets information regarding upstream process changes that influence influent hydraulics to the UV system, such as status of pump stations or filter backwash processes. In some instances, the facility SCADA system may be tasked with active control of the UV disinfection system.

Ultraviolet system statuses, warnings, and alarms may be locally annunciated at the UV system control cabinet(s) and/or be transmitted to the facility SCADA system for operation staff monitoring, historical logging, and remote annunciation. The SCADA integration can allow significant/

critical alarms to be connected to a remote annunciation system (such as an auto-dialer) to notify off-site staff of the alarm. Some systems may be configured so that facility SCADA systems will automatically shut down UV equipment upon certain critical UV alarms or conditions outside of the UV system scope.

Facility SCADA systems may include historical logging and trending features. Data collected by the master PLCs, such as UV lamp statuses, elapsed time meters, warnings, and alarms, can be stored in the SCADA system or an external database. Trending and reporting features can be set up to fulfill the needs of operations staff. In particular, trending UV dose, UV intensity, lamp life, and power consumption can be helpful to evaluate UV system performance over time. Trending effluent flow and UVT is strongly recommended on an ongoing basis because it can be used to reassess capacity of the UV facility at any point in time. Trending can also be used to estimate maintenance costs or validate warranty claims.

Ultraviolet disinfection systems are typically provided with one control cabinet per UV reactor. In systems with multiple duty banks or reactors, the control boards/PLCs in each cabinet must communicate with each other to implement all but the simplest manual control schemes. For example, for flow-pacing or dose-pacing control schemes, the control board/PLC for each reactor must know how many lamps are in service in the other reactors. This communication can happen in the following different ways:

- A master control panel provided by the UV system manufacturer, communicating with each control cabinet;
- Direct communication between each control cabinet, with the control board/PLC in one control cabinet designated as the master; and
- The control cabinets by the UV system manufacturer each communicate with the facility SCADA system or a facility-provided master control panel, where the UV system flow- or dose-pacing algorithms are resident. These algorithms may be provided by the UV system manufacturer or a system integrator retained by the treatment facility.

Facility staff should always be aware of the system failure modes in case of loss of communication. For example, how will the UV control system react to a loss of the effluent flow meter signal, either because of instrument failure or a SCADA system failure? Failure modes must be programmed into each control cabinet for loss of all external data sources and documented in the O&M manual.

6.0 MAINTENANCE CONSIDERATIONS

Proper maintenance of any equipment system is required to provide efficient operations, including the following (U.S. EPA, 2006; WEF, 2006):

- Weekly—check and calibrate the online UVT analyzers;
- Monthly—check the quartz sleeves and wipers for cracks and/or leaks;
- Weekly to monthly—check the cleaning efficacy. If the UV system is not equipped with an automatic wiping system, the quartz sleeves must be cleaned manually, which can range from simple wiping by hand to in-channel cleaning or a separate tank dedicated to cleaning;
- Quarterly—test the ground fault interrupt;
- Semiannually—check the cleaning fluid reservoir (semiannually);
- When required
  - Check the accuracy of the flow meters (as recommended by manufacturer);
  - Replace lamps
    - When a lamp fails,
    - Reaches the predetermined service hours,
    - The system does not achieve the required performance,
    - Online monitoring alarm signals lamp replacement, and
    - The UV monitor indicates that the UV intensity is low;
  - Replace quartz sleeves
    - At prespecified schedule,
    - When damage or cracks are observed, and
    - When excessive fouling occurs that cannot be removed by cleaning; and
  - Replace ballasts when a ballast fails.

Table 8.3 summarizes significant maintenance considerations for UV systems.

6.1 Lamp Replacement

Lamps are replaced at a frequency that typically ranges from 6 months to 3 years. Some facilities replace lamps strictly in accordance with manufacturer's recommendations, while other facilities only replace them when intensity decreases or indicator counts increase.

TABLE 8.3 Typical maintenance tasks.

Frequency	Task or general guideline	Action
Weekly	Check online UV transmittance analyzer calibration (if applicable).	Calibrate UV transmittance analyzer when manufacturer's guaranteed measurement uncertainty is exceeded.
Monthly	Check sleeves and wipers for leaks.	Replace sleeves, seats, O-rings, or wiper seals if damaged or leaking.
Monthly	UV intensity calibration check.	Check sensor calibration at the lamp power output used during routine operating conditions.
Monthly	Check cleaning efficiency.	Record UV intensity sensor reading. Examine sleeves for streaks or remaining deposits. Record UV intensity sensor reading after cleaning and compare with precleaned value. Reclean or replace sleeves if intensity is not restored to precleaned value.
Semiannually	Check cleaning fluid reservoir (if applicable).	Replenish solution if the reservoir is low. Drain and replace solution if it is discolored.
Annually	Test trip ground fault interrupt (GFI).	Maintain GFI breakers in accordance with manufacturers' recommendations.
Plant-specific	Replace lamps.	Replace lamps when any of the following conditions occur: • Initiation of low UV intensity after verifying that condition is caused by low lamp output, • Initiation of lamp failure alarm, and • System not achieving permit requirements after exceeding guaranteed lamp run time.
Manufacturer's recommended frequency	Check flow meter calibration.	If effluent weir or flume is available, manually check depth and compare with primary device measurement. Primary device should be calibrated at recommended frequency or when measurement uncertainty is exceeded.
	Clean and calibrate transmittance monitor (if installaled).	Clean and calibrate according to manufacturer's recommended frequency.
When lamps are replaced	Properly dispose of lamps.	Send spent lamps to a mercury recycling facility or back to the manufacturer.
When quartz sleeves are replaced	Properly dispose of quartz sleeves.	Replace sleeves as recommended by the manufacturer or when damage, cracks, or excessive fouling occurs that would impede UV intensity.

6.2 Sleeve Replacement

Similar to lamps, sleeves are replaced at a frequency that typically ranges from 1 to 5 years. Similar to lamps, some facilities only replace sleeves when intensity decreases or indicator counts increase and, for low-pressure systems, they may retain sleeves in use for the life of the system.

6.3 Ballast Replacement

Ballasts are replaced at a frequency that typically ranges from 3 to 8 years.

6.4 Sleeve Cleaning

Control of lamp sleeve fouling is achieved by a variety of techniques. The following are sleeve-cleaning options that are used currently:

- Manual cleaning strategies, which require periodic removal of the sleeves for soaking in a chemical bath or manual wiping with a chemical cleaner;
- Automated online strategies, which use mechanical cleaning devices that wipe frequently and require periodic manual chemical cleaning;
- Automated chemical/mechanical cleaning systems (that may require periodic manual chemical cleaning [e.g., for area's outside the lamp's arc length and/or depending on the nature of scaling constituents]);
- Chemical removal of scale is achieved by applying a dilute acid (pH of approximately 1 to 3) to the fouled surface. Acid can be applied by either wiping individual lamps or immersing entire lamp modules. Immersion techniques probably are more efficient for scale removal. For large systems, module-immersion hardware is a necessity;
- Several different acid solutions have been used for chemical cleaning, including citric acid, phosphoric acid, and commercially available bathroom cleaners. Selection of an appropriate acid will depend on site-specific requirements, but disposal of spent acids should be incorporated to the decision. For large systems, the use of food-grade citric acid or phosphoric acid should be considered so that the neutralized liquid containing the spent acid can be diverted to the headworks of the WRRF (it is important to note that operators should first consider the effect of recycle streams containing spent cleaning solutions on facility performance);
- A number of physical processes can be incorporated to mitigate scaling. Introducing air bubbles at the base of a channel for short periods of time, but frequently (e.g., 10 min/d), has been performed

successfully at some UV systems. This procedure will not eliminate the need for cleaning at facilities where fouling occurs, but will be effective in increasing the interval between cleanings;

- Citric acid, phosphoric acid, proprietary mixtures, and commercially available bathroom cleaners were used most commonly to clean sleeves. Typically, WRRFs should use a commercially available, inexpensive cleaning agent that is handled and disposed of easily. Materials issues (e.g., corrosion) should be considered if cleaning is to be performed in situ. A small, bench-scale, flow-through unit can be used to evaluate a number of agents by trial and error and the optimal cleaning frequency. Historical reported manual cleaning frequencies are highly site-specific and range from weekly to yearly, with a median frequency of approximately once per month (U.S. EPA, 1992). However, with the proliferation of automated cleaning systems, cleaning cycle times have increased from these levels;

- Low-pressure, low-intensity lamp cleaning in horizontal systems is typically accomplished by either bank or module removal to a mobile or dedicated cleaning station. The level of cleaning complexity can range from a drained area equipped with a holding rack, hose, and cleaning solution to automatic air sparging or an ultrasonic dip tank for large banks accessed with overhead hoists; and

- Low-pressure, low-intensity vertical lamp cleaning typically is accomplished in a similar manner to that of horizontal systems. Current options include dip tanks and an air-scouring system, which is engaged in place and under process conditions. It is used to increase the interval between chemical lamp-cleaning cycles, which can either be done in situ (isolating the channel) or by transferring the module to a dip tank.

6.5 Channel Cleaning

The deposition of solids and growth of algae are known to result in permit exceedances because both solids and algae can harbor bacteria. Cleaning these nuisance items is standard practice for successful operation of UV systems. Solids accumulation and algae growth will occur in most UV reactors, with tertiary filtration reducing the amount of solids deposition. Solids accumulation and growth will also occur on the UV modules, outside of the arc length of the lamps; growth has also been observed downstream of the UV systems on the effluent weirs. Thus, to minimize the effect of solids and algae, the UV channels (and pressurized UV reactors) should be cleaned on a schedule. While cleaning appears to be primarily an issue for O&M,

the design team must recognize the effect of solids deposition and maintenance activities. Suggested design features to facilitate cleaning include the following:

- Cover the channel to prevent sunlight from causing algae growth and to prevent debris and other solid material from falling into the reactor;
- Use redundancy in multiple channel/train designs to allow one of the channels/trains to allow one channel to be routinely be taken out of service and cleaned;
- Reactors, channels, and related tankage should be equipped with drains to allow complete and rapid dewatering; drainage should be directed back to the headworks of the WRRF; and
- A clean water system should be permanently available for rinsing and cleaning needs.

Some UV designs also incorporate lining or coating of the channel, minimizing surfaces where biofilms can attach. This can be accomplished by either lining the channel with stainless steel or coating the channel with a UV-resistant material. The material that should be used in the lining or coating of the process should be able to withstand a high-pressure spray wash that may be required to remove the biological growths that will occur during normal operation.

7.0 ANCILLARY SYSTEMS

To support operations of the significant equipment that is installed to provide disinfection, there are a number of ancillary systems that must also be maintained. A discussion of considerations for O&M of some of the key ancillary systems is provided in the following sections.

7.1 Chemical Cleaning Tanks/Stations

Various cleaning methods are provided with current UV disinfection systems. For small open-channel systems, manual chemical cleaning and wiping may be sufficient. Cleaning racks for individual lamps or lamp modules should be provided in a drained area. In larger open-channel systems, lamp modules or entire banks may be removed at one time and cleaned in a dip tank. In this case, a hoist or crane is needed for removing and handling the banks.

At larger facilities, chemical storage tanks and metering pumps may be required for supply of cleaning chemicals. Storage and safety precautions should be in accordance with the chemical supplier's recommendations.

7.2 Air Compressors/Blowers

Some open-channel UV systems are equipped with an air-scouring system that is engaged in place in the UV channels and under process conditions. The air scour increases the interval between chemical lamp cleaning cycles.

Some UV systems with mechanical wiping systems use air compressors for pneumatic control of the wipers.

In both of these cases, the UV system requires a pressurized air supply. The UV system supplier may provide an air blower or compressor as part of their scope of supply. Alternatively, air for the UV system could be provided from the facility air system if the air quality and quantity meet the requirements of the UV system supplier.

7.3 Bank/Module Lifting

In larger open-channel systems, a hoist or crane may be required for removing and handling the banks. Entire banks may be removed at one time and cleaned in a dip tank. Bank removal, or at least module removal, may also be required for lamp inspection and replacement.

Jib cranes or davit hoists can be mounted adjacent to the UV channels. To transfer a bank to a cleaning tank, a monorail hoist system or a bridge crane could also be used. In planning a new crane installation, consider turning radius for crane access. Various types of cranes should be considered and reviewed for UV systems.

8.0 TROUBLESHOOTING

Table 8.4 summarizes typical troubleshooting considerations for UV systems. Additional discussion on common areas that require troubleshooting is provided in the following sections.

8.1 Electrical Issues

Issues with power and control circuits, including for the ballasts, lamps, and control systems, are common causes of UV problems. It is typically recommended that manufacturer's manuals be consulted for troubleshooting electrical issues. Proper precautions are essential; often, electrical troubleshooting requires a licensed electrician.

8.2 Low Ultraviolet Transmittance

Critical data to be evaluated during the design phase include flow, UVT, suspended solids, and viable indicator organism (e.g., coliform bacteria)

TABLE 8.4 Troubleshooting checklist.

Equipment	What to check	Potential problems	Corrective actions
Ballasts	Surface temperature on ballasts during operation on normal utility power	Overheating resulting from poor panel ventilation	Add panel ventilation or cooling system.
	Surface temperature on ballasts during operation on standby power	Overheating resulting from power-distorting harmonics from electronic ballasts	Check power quality under various UV loads. May require addition of system or equipment to filter out harmonics.
	Proper grounding	Frequent ballasts failures	Adequate grounding per UV equipment supplier recommendations.
UV system	Lamp indicators	Burned out bulb	Replace as needed.
		Wrong sequence	Respective lamps indicate sequence of individual components.
Intensity meter	Indicates chamber UV intensity	Buildup on quartz jacket	Clean routinely as necessary.
Intensity monitor	Photocell and electronic circuit, indicating meter alarm condition and pilot lights	Nonfunctional	Repair or replace.
UV lamps	Lamps	Burned out	Replace as necessary.
	Heat buildup or LP lamp exposure to air	Little or no flow	Increase water supply.
		Malfunction of level control system	Clean level sensors; repair or replace as necessary.
	GFI indicator	Broken quartz sleeve or seal system failure	Check sleeve(s) for breaks and leaks; dry contacts; replace lamp system components as required.
Control box indicator lights	Amber	Low UV output	Clean chamber or replace bulb.
	Red	Poor water quality	Clean chamber or replace bulb. Check processes.
Gland seal assembly	O-ring	Water leaks	Tighten the gland nut to compress O-ring, or replace.
Electrical service	AC volts, DC volts, ohms AC	Over range	Use multimeter and set ranges according to manufacturer's recommendation.
Lamp out warning system	Circuit board	Defective	Replace.
	Indicator bulb	Burned out	Replace.
	Pilot light	Burned out	Replace.

concentrations. The UVT data, trended through the SCADA historian, should be analyzed for low UVT trends. The timing of these trends can be used to track down industries, or other sources, which may be contributing low UVT flow to the WRRF. Typically, the UVT of secondary effluent will be greater than 60% after the effluent is filtered through a 0.45-micron filter, although lower values (approximately 50%) have been observed. In some instances, industrial effluents may contribute components with high UV absorbance, which can strongly affect the UVT of the combined domestic and industrial wastewater. When collecting UVT data, especially when troubleshooting low UVT, it is important to measure the parameter on both filtered and unfiltered samples to determine if dissolved constituents may be absorbing UV light (Swift et al., 2001). In some instances, particularly at low transmittance levels, it may be necessary to reduce the spacing of the lamps or consider using advanced higher power low-pressure high-output systems to overcome the lower transmittance of the water.

A number of industries have been implicated as discharging wastewater with high dissolved UV absorbance (low filtered UVT) as a result of the presence of organic compounds not readily degraded (Swift et al., 2007), including sunblock, coffee, pharmaceutical, and chemical manufacturers; centralized waste treatment facilities; and printed-circuit-board manufacturers. Several WRRFs have been documented as having potentially violated effluent disinfection standards as a result of the presence of refractory organic compounds passing through the WRRF and lowering effluent UVT. Compounds in the wastewater responsible for lowering effluent UVT can be considered to be "pollutants of concern" for source control programs and thus appropriate for regulatory control through the use of local limits established through an evaluation of maximum allowable headworks loading using methods analogous to those used for other pollutants (Swift et al., 2007).

8.3 Hydraulics

Hydraulic issues are fairly common causes of problems with UV systems, particularly when upstream or downstream equipment or processes have changed. A common problem with open-channel systems is when the water level is too high or too low, as might occur if a level control gate is malfunctioning. Checking water level is often recommended when disinfection performance has decreased and problems with lamps, ballasts, or UVT have been ruled out.

8.4 Upstream Process Effects

Upstream treatment processes have a significant effect on the disinfectability of secondary effluent. Because of "shading", particle-associated coliforms

(PACs) often are difficult to inactivate to levels required for reuse (e.g., 2.2 total coliforms). Additionally, certain upstream processes, such as fixed-film biological treatment, yield recalcitrant effluent with a high concentration of PACs in large particles. Loge et al. (1999) reported that PACs were found to decline exponentially with increasing mean cell residence times (MCRTs). The factors influencing the formation of PACs included the concentration of particles, concentration of dispersed (non-particle-associated) coliform bacteria, and the MCRT. The concentration of dispersed coliform bacteria was found to decline with increasing MCRTs. The rate of decline was greater than the typical half-life attributed to endogenous decay, suggesting that other factors (e.g., predation by protozoa) influence the concentration of dispersed coliform bacteria and, subsequently, the formation of PAC (Loge et al., 1999).

9.0 REFERENCES

Loge, F. J.; Emerick, R. W.; Thompson, D. E.; Nelson, D. C; Darby, J. L. (1999) Factors Influencing Ultraviolet Disinfection Performance Part I: Light Penetration to Wastewater Particles. *Water Environ. Res.* **71** (3), 377–381.

National Water Research Institute; Water Research Foundation (2012) *Ultraviolet Disinfection Guidelines for Drinking Water and Water Reuse*, 3rd ed.; National Water Research Institute: Fountain Valley, California.

Swift, J. L.; Wilson, J. P.; Hunter, G. (2007) Implementing Local Limits for the Control of WWTP Effluent Ultraviolet Transmittance. *Proceedings of the 80th Annual Water Environment Federation Technical Exposition and Conference* [CD-ROM]; San Diego, California, Oct 13–17; Water Environment Federation: Alexandria, Virginia.

Swift, J. L.; Wilson, J. P.; Johnson, M.; Jacobsen, B. (2001) The Impact of UV-Absorbing Wastewater from Printed Circuit Board Manufacturing Facility on the Performance of a Municipal UV Disinfection System. *Proceedings of the 74th Annual Water Environment Federation Technical Exhibition and Conference* [CD-ROM]; Atlanta, Georgia, Oct 13–17; Water Environment Federation: Alexandria, Virginia.

U.S. Environmental Protection Agency (2006) *Ultraviolet Disinfection Guidance Manual for the Final Long Term 2 Enhanced Surface Water Treatment Rule*; EPA-815/R-06-007; U.S. Environmental Protection Agency: Washington, D.C.

U.S. Environmental Protection Agency (1992) *Ultraviolet Disinfection Technology Assessment*; EPA-832/R-92-004; U.S. Environmental Protection Agency: Washington, D.C.

Water Environment Federation (2012) *Safety, Health, and Security in Wastewater Systems*, 6th ed.; WEF Manual of Practice No. 1; Water Environment Federation: Alexandria, Virginia.

Water Environment Federation (2006) *Wastewater Disinfection Training Manual*; Water Environment Federation: Alexandria, Virginia.

Water Environment Research Foundation (2007) *Disinfection of Wastewater Effluent—Comparison of Alternative Technologies;* Report No. 04-HHE-4; Water Environment Research Foundation: Alexandria, Virginia.

9

Case Studies

Oliver Lawal and Bree Trembly

1.0	OVERVIEW	191	4.0 CLOSED VESSEL	201
2.0	OPEN CHANNEL, HORIZONTAL ORIENTATION	191	5.0 REFERENCE	204
3.0	OPEN CHANNEL, NONHORIZONTAL ORIENTATION	197		

1.0 OVERVIEW

Case study examples in this chapter present the diversity of UV technologies, hardware configurations, and design parameters currently deployed by a broad range of manufacturers. The selection reflects the broad range of configurations seen both within the United States and selected locations globally. A summary of the case studies is provided in Table 9.1. Design principles discussed in previous chapters have been applied in these examples.

2.0 OPEN CHANNEL, HORIZONTAL ORIENTATION

The most common configuration for wastewater UV system deployment is within open concrete channel(s). To limit civil construction works, it is common practice to repurpose former chlorine retention channels for use with UV systems. However, purpose-built concrete and stainless steel channels are also seen. Tables 9.2 to 9.5 show examples of UV installations with the lamps positioned horizontally within open channels, which is historically the most common configuration. Concerns regarding level control reliability and hydraulic inefficiencies cause some to favor configurations with nonhorizontal lamps and others to favor a closed-pipe configuration.

TABLE 9.1 Summary of UV system design case study examples; case study examples represent a variety of systems that have been procured and constructed using the various methods described in the previous chapters (mgd ÷ 0.2642 = ML/d).

Chapter section	Site, location	Manufacturer	Configuration	Lamp type	Lamp driver	Number of lamps	Maximum flowrate
2.1	Owen, Wisconsin	Trojan	Open channel, horizontal	LPLO[a]	Electronic	48	1 mgd
2.2	Napoleon, Ohio	Enaqua	Open channel, horizontal	LPHO[a]	Electronic	792	20 mgd
2.3	Manukau, New Zealand	Wedeco	Open channel, horizontal	LPHO[a]	Electronic	7776	376 mgd
2.4	Penn Hills/Plum Creek	Trojan	Open channel, horizontal	Medium-pressure	Magnetic	168	12 mgd
3.1	Jefferson, Missouri	IDI/Ozonia	Open channel, vertical	LPHO[a]	Electronic	648	66.6 mgd
3.2	Auburn, Alabama	Trojan	Open channel, inclined	LPHO[a]	Electronic	116	34 mgd
3.3	Rensselaer, New York	Wedeco	Open channel, inclined	LPHO[a]	Electronic	216	64 mgd
3.4	Macon, Missouri	Severn Trent	Open channel, vertical	LPHO[a]	Microwave	96	5.4 mgd
4.1	Sieldlce, Poland	Berson	Closed vessel, horizontal	Medium-pressure	Electronic	24	19 mgd
4.2	Horizon, Texas	ETS	Closed vessel, horizontal	LPHO[b]	Electronic	64	9 mgd
4.3	Gainesville, Georgia	Aquionics	Closed vessel, horizontal	Medium-pressure	Electronic	96	24 mgd

[a]Low-pressure low-output.
[b]Low-pressure high-output.

TABLE 9.2 Summary of information on the Owen, Wisconsin, water resource recovery facility, representing an open-channel system with lamps oriented in a horizontal configuration (information submitted by Trojan Technologies).

Site background	Location	Owen, Wisconsin
	Facility process type	Activated sludge
	Use	River discharge
	Effluent limit	400 cfu/100 mL
	Procurement method	Direct purchase by owner
Key design parameters	Design dose	30 mJ/cm^2 reduction equivalent dose MS2
	Design method	U.S. EPA, *Design Guidance Manual*
	Design (peak) flow	157.6 m^3/h (1 mgd)
	Average flow	74 m^3/h (0.47 mgd)
	UV transmittance	55%
	Total suspended solids	30 mg/L
Equipment	Manufacturer	Trojan Technologies
	Model	UV3000
	Type	Open channel—horizontal lamps
	Configuration	2 channels; 6 modules/channel
	Total number of lamps	48
	Redundancy	none
	Power consumption per mgd	4.2 kW
	Headloss at design flow	1.4 cm/bank (0.55 in.)
Installation and commissioning	Year commissioned	2013
	Performance testing	Fecal coliform
	Installation location	Indoor
Operational and control strategy	Control strategy	Flow pacing
	Ballast turndown range	100% (no turndown)
	Level control method	Fixed weir
Operation and maintenance	Annual operation	365 d/yr
	Lamp replacement period	12 000 hours
	Ballast replacement period	7 years
	Sleeve cleaning method	Manual

TABLE 9.3 Summary of information on the Napolean, Ohio, water resource recovery facility, representing an open-channel system with lamps oriented in a horizontal configuration (information and photograph submitted by Enaqua) (mgd ÷ 0.2642 = ML/d).

Picture		
Site background	Location	Napoleon, Ohio
	Facility process type	Combined sewer overflow
	Use	River discharge
	Effluent limit	200 cfu/100 mL
	Procurement method	Not available
Key design parameters	Design dose	32.8 mJ/cm^2, average
	Design method	Proprietary sizing method
	Design (peak) flow	3152 m^3/h (20 mgd)
	Average flow	1182 m^3/h (7.5 mgd)
	UV transmittance	40% based on sampling results
	Total suspended solids	30 mg/L
Equipment	Manufacturer	Enaqua
	Model	C-Series
	Type	Open channel, flow through—horizontal lamps
	Configuration	2 channels, 3 banks/channel
	Total number of lamps	792/132 per bank
	Redundancy	None
	Power consumption	5.7 kW/mgd
	Headloss at design flow	26.4 cm (10.4 in.) @ 7.5 mgd
Installation and commissioning	Year commissioned	2010
	Performance testing	*E. coli*
	Installation location	Outdoor, canopy cover
Operational and control strategy	Control strategy	Dose control by bank switching, level pacing. Based on intensity and flowrate and variable channel level
	Ballast turndown range	Level based—on/off
	Level control method	Natural orifice action
Operation and maintenance	Annual operation	~5760 hours/year
	Lamp replacement period	16 000 hours
	Ballast replacement period	None replaced yet; guarantee 5 years
	Sleeve cleaning method	AFP™ Tubing—NO quartz high velocity self-cleaning—no mechanical cleaning mechanism

TABLE 9.4 Summary of information on the Manukau water resource recovery facility in Auckland, New Zealand, representing an open-channel system with lamps oriented in a horizontal configuration (information and photograph submitted by Xylem).

Picture		
Site background	Location	Auckland, New Zealand
	Facility process type	Tertiary, combined sewer overflow blend
	Use	Coastal discharge
	Effluent limit	2-log bacteriophage, 4-log pathogen
	Procurement method	Consultant predesign/selection
Key design parameters	Design dose	32.8 mJ/cm² reduction equivalent dose
	Design method	On-site bioassay
	Design (peak) flow	59 267 m³/h (376 mgd)
	Average flow	12 452 m³/h (79 mgd)
	UV transmittance	55%
	Total suspended solids	15 mg/L
Equipment	Manufacturer	Wedeco—Xylem
	Model	TAK 55 9-12 × 3i12W
	Type	Open channel—horizontal lamps
	Configuration	12 channels, 3 banks/channel, 12 modules/bank, 18 lamps/module
	Total number of lamps	7776
	Redundancy	3 of 12 channels reserved for storm-water ingress (weather events)
	Power consumption	Maximum 2.56 MW
	Headloss at design flow	Not available
Installation and commissioning	Year commissioned	2001
	Performance testing	Fecal coliform at commissioning and annual
	Installation location	Indoor
Operational and control strategy	Control strategy	Dose control by bank switching and lamp output. Based on intensity, flowrate, and UVT input
	Ballast turndown range	50–100%
	Level control method	Motorized weir gate
Operation and maintenance	Annual operation	365 d/yr
	Lamp replacement period	12 000 hours
	Ballast replacement period	10 years
	Sleeve cleaning method	Pneumatic-driven mechanical wiping

TABLE 9.5 Summary of information on the Penn Hills/Plum Creek water resource recovery facility, Pennsylvania, representing an open-channel system with lamps oriented in a horizontal configuration (information and photograph submitted by Trojan Technologies).

Picture		
Site background	Location	Plum Creek, Pennsylvania
	Facility process type	Conventional activated sludge
	Use	River discharge
	Effluent limit	80 cfu/100 mL
	Procurement method	Public contractor bid
Key design parameters	Design dose	20 mJ/cm^2 determined by collimated beam study
	Design method	National Water Research Institute (NWRI) bioassay
	Design (peak) flow	1891 m^3/h (12 mgd)
	Average flow	4164 m^3/h (6 mgd)
	UV transmittance	65%
	Total suspended solids	10 mg/L
Equipment	Manufacturer	Trojan Technologies
	Model	UV 4000+
	Type	Open channel—horizonal medium-pressure lamps
	Configuration	2 channels, 6 modules/channel, 14 lamps/module
	Total number of lamps	168
	Redundancy	None
	Power consumption per mgd	Not available
	Headloss at design flow	Not available
Installation and commissioning	Year commissioned	2011
	Performance testing	Onsite fecal coliform bioassay
	Installation location	Outdoor, covered
Operational and control strategy	Control strategy	Dose control by variable lamp output; switching banks on/off by flow pacing
	Ballast turndown range	30–100%
	Level control method	Motorized weir gate
Operation and maintenance	Annual operation	365 d/yr
	Lamp replacement eriod	5000 hours
	Ballast replacement period	7 years
	Sleeve cleaning method	Motor-driven, chemical and mechanical wiping

3.0 OPEN CHANNEL, NONHORIZONTAL ORIENTATION

Tables 9.6 to 9.9 show examples of UV installations with the lamps positioned in a vertical, or inclined (nonhorizontal) configuration within an open channel or open channels. The channel is typically deeper with these configurations and allows the use of longer lamps, which some manufacturers claim to be more efficient (see Table 9.7). A number of manufacturers also claim that nonhorizontal configurations allow for more efficient hydraulic flow though the lamps. Concerns regarding level control reliability, resulting

TABLE 9.6 Summary of information on the City of Jefferson, Missouri, Regional Water Reclamation Facility, representing an open-channel system with lamps oriented in a vertical configuration (information submitted by CDM Smith) (mgd ÷ 0.2642 = ML/d).

Site background	Location	Jefferson, Missouri
	Facility process type	Continuously fed sequencing batch reactor
	Use	River discharge
	Effluent limit	30-day geo mean of 206 cfu/100 mL for *E. coli*
	Procurement method	Not available
Key design parameters	Design dose	30 mJ/cm^2
	Design method	NWRI and Water Research Foundation (WRF)
	Design (peak) flow	10 498 m^3/h (66.6 mgd)
	Average flow	1797 m^3/h (11.4 mgd)
	UV transmittance	65%
	Total suspended solids	Monthly average = 30 mg/L; 7-day average = 45 mg/L
Equipment	Manufacturer	IDI/Ozonia
	Model	Aquaray
	Type	Open channel—vertical lamps
	Configuration	2 channels, 3 banks/channel, 3 modules/bank, 36 lamps/module
	Total number of lamps	648 (432 duty)
	Redundancy	216 lamps
	Power consumption per mgd	2.69 kW
	Headloss at design flow	22.8 cm (9 in.) @ 33.3 mgd per channel
Installation and commissioning	Year commissioned	2011
	Performance testing	3-day performance testing, including 1 day of total suspended solids stress testing
	Installation location	Outdoor covered, electrical indoor
Operational and control strategy	Control strategy	Flow pacing
	Ballast turndown range	50–100%
	Level control method	Fixed finger weir
Operation and maintenance	Annual operation	5136 hr/yr
	Lamp replacement period	9000–12 000 hours
	Ballast replacement period	Every 5 years
	Sleeve cleaning method	Motor-driven, mechanical wiping with air scour

TABLE 9.7 Summary of information on the Auburn water resource recovery facility, representing an open-channel system with lamps oriented in an inclined configuration (information and photograph submitted by Trojan Technologies) (mgd ÷ 0.2642 = ML/d).

Picture		
Site background	Location	Auburn, Alabama
	Facility process type	Conventional activated sludge
	Use	River discharge
	Effluent limit	126 cfu/100 mL *E. coli*
	Procurement method	Public contractor bid
Key design parameters	Design dose	15 mJ/cm^2 T1 phage, reduction equivalent dose
	Design method	NWRI and WRF, U.S. EPA *Design Guidance Manual*
	Design (peak) flow	5391 m^3/h (34.2 mgd)
	Average flow	1418 m^3/h (9 mgd)
	UV transmittance	65%
	Total suspended solids	30 mg/L
Equipment	Manufacturer	Trojan Technologies
	Model	Signa
	Type	Open channel—inclined lamps
	Configuration	2 channels, 2 modules/channel, 29 lamps/module
	Total number of lamps	116
	Redundancy	None
	Power consumption	3.4 kW/mgd
	Headloss at design flow	7.11 cm (2.8 in.)
Installation and commissioning	Year commissioned	2013
	Performance testing	Ongoing
	Installation location	Outdoor, canopy covered
Operational and control strategy	Control strategy	Dose control by bank switching and lamp output based on flowrate and UVT
	Ballast turndown range	30–100%
	Level control method	Motorized weir gate
Operation and maintenance	Annual operation	365 d/yr
	Lamp replacement period	15 000 hours
	Ballast replacement period	10 years
	Sleeve cleaning method	Motor-driven, chemical and mechanical wiping

TABLE 9.8 Summary of information on the Rensselaer water resource recovery facility, representing an open-channel system with lamps oriented in an inclined configuration (information and photograph submitted by Xylem) (mgd ÷ 0.2642 = ML/d).

Picture		
Site background	Location	Rensselaer, New York
	Facility process Type	Secondary clarifier
	Use	(Hudson) River discharge
	Effluent limit	200 cfu/100 mL
	Procurement method	Public contractor bid
Key design parameters	Design dose	30 mJ/cm^2 average dose
	Design method	MS-2 bioassay, "IUVA protocol"
	Design (peak) flow	10 088 m^3/h (64 mgd)
	Average flow	3782 m^3/h (24 mgd)
	UV transmittance	65%
	Total suspended solids	<30 mg/L
Equipment	Manufacturer	Wedeco–Xylem
	Model	Duron
	Type	Open channel—inclined lamps
	Configuration	2 channels, 9 modules/channel, 12 lamps/module
	Total number of lamps	216
	Redundancy	None at design flow
	Power consumption	2.5 kW/mgd lamps and ballasts only
	Headloss at design flow	24.13 cm (9.5 in.)
Installation and commissioning	Year commissioned	2013
	Performance testing	None
	Installation location	Outdoor, exposed
Operational and control strategy	Control strategy	Intensity control by bank switching and lamp output. Based on intensity, flowrate, and UVT input
	Ballast turndown range	50–100%
	Level control method	Mechanically driven downward opening gates
Operation and maintenance	Annual operation	Seasonal—May to Oct
	Lamp replacement period	14 000 hours
	Ballast replacement period	Guarantee 5 years
	Sleeve cleaning method	Motor-driven mechanical wiping

TABLE 9.9 Summary of information on the Macon, Missouri, water resource recovery facility, representing an open-channel system with microwave lamps oriented in a vertical configuration (information submitted by CDM Smith) (mgd ÷ 0.2642 = ML/d).

Site background	Location	Macon, Missouri
	Facility process type	Two-stage trickling filter
	Use	River discharge
	Effluent limit	<206/100 mL based on 30-day geometric mean
	Procurement method	Competitive preselection
Key design parameters	Design dose	15 mJ/cm^2 determined by multiple collimated beam studies
	Design method	T1 bioassay, "IUVA protocol"
	Design (peak) flow	851 m^3/h (5.4 mgd)
	Average flow	315 m^3/h (2.0 mgd)
	UV transmittance	60%
	Total suspended solids	30 mg/L
Equipment	Manufacturer	Severn Trent
	Model	MicroDynamics
	Type	Open channel—horizontal lamps
	Configuration	2 channels, 3 modules/channel, 16 lamps/module
	Total number of lamps	96
	Redundancy	20%
	Power consumption	7.2 kW/mgd
	Headloss at design flow	5.72 cm (2.25 in.)
Installation and commissioning	Year commissioned	2011
	Performance testing	2 days normal operation, 1 day of stress-testing with elevated total suspended solids levels
	Installation location	Outdoor, canopy covered
Operational and control strategy	Control strategy	Dose control, flow pacing
	Ballast (magnetron) turndown range	80–100%
	Level control method	Fixed finger weir
Operation and maintenance	Annual operation	5136 hours
	Lamp replacement period	24 000 to 27 000 hours
	Ballast (magnetron) replacement period	2–3 years
	Sleeve cleaning method	Mechanical wipers

in a top layer of untreated flow, or potential exposure of the lamps to air (causing overheating) still persist with these configurations.

4.0 CLOSED VESSEL

Closed vessel configurations, shown in Tables 9.10 to 9.12 have the lamps enclosed within a stainless steel reactor. This allows for a smaller footprint

TABLE 9.10 Summary of information on Sieldlce Waterworks, Poland, representing a closed-vessel system (information submitted by Berson UV-techniek) (mgd ÷ 0.2642 = ML/d).

Site background	Location	City of Sieldlce, Poland
	Facility process type	Activated sludge
	Use	River discharge (on recreation water)
	Effluent limit	Total coliforms ≤200 cfu/100 mL
	Procurement method	Consultant predesign
Key design parameters	Design dose	40 mJ/cm^2 average dose (20 mJ/cm^2 contribution from each unit in series)
	Design method	Computational fluid dynamics modeling and bioassay testing
	Design (peak) flow	3000 m^3/h (19 mgd)
	Average flow	1875 m^3/h (11.8 mgd)
	UV transmittance	Min = 60%, average = 65%
	Total suspended solids	<20 mg/L
Equipment	Manufacturer	Berson
	Model	InLine 33000 + WW Summit
	Type	Closed vessel—horizontal, perpendicular lamps
	Configuration	2 systems positioned in series, configuration = duty + assist
	Total number of lamps	24
	Redundancy	75% redundancy @ average design flow
	Power consumption per mgd	Approx. 8.0 kW/mgd
	Headloss at design flow	63.5 cm (25 in.)
Installation and commissioning	Year commissioned	2013
	Performance testing	Fecal coliform at commissioning
	Installation location	Indoors, in basement
Operational and control strategy	Control strategy	Dose pacing
	Ballast turndown range	33–100%
	Level control method	Not applicable—pressurized vessel
Operation and maintenance	Annual operation	365 d/yr
	Lamp replacement period	Expected 9000 hours
	Ballast replacement period	Expected >10 years
	Sleeve cleaning method	Motor-driven automatic mechanical

TABLE 9.11 Summary of information on the Horizon, Texas, water resource recovery facility, representing a closed-vessel system (information and photo submitted by ETS) (mgd ÷ 0.2642 = ML/d).

Picture		
Site background	Location	Horizon, Texas
	Facility process type	Tertiary
	Use	River discharge
	Effluent limit	2000 FC/100 mL
	Procurement method	Public contractor bid
Key design parameters	Design dose	30 mJ/cm^2 MS-2 reduction equivalent dose
	Design method	NWRI and WRF
	Design (peak) flow	1419 m^3/h (9 mgd)
	Average flow	551 m^3/h (3.5 mgd)
	UV transmittance	75%
	Total suspended solids	<10 mg/L
Equipment	Manufacturer	Engineered treatment systems
	Model	UVLW-16800-20
	Type	Closed vessel—horizontal, inline lamps
	Configuration	4 vessels, 16 lamps/vessel
	Total number of lamps	64
	Redundancy	33.3%
	Power consumption per mgd	23.26 kW/mgd
	Headloss at design flow	12.7 cm (5 in.)
Installation and commissioning	Year commissioned	2011
	Performance testing	Daily sampling
	Installation location	Outdoor, canopy covered
Operational and control strategy	Control strategy	Dose pacing
	Ballast turndown range	30–100%
	Level control method	N/A—pressurized vessel
Operation and maintenance	Annual operation	365 d/yr
	Lamp replacement period	14 000 hours
	Ballast replacement period	43 000 hours
	Sleeve cleaning method	Motor-driven automatic mechanical

TABLE 9.12 Summary of information on the Flat Creek water resource recovery facility, representing a closed-vessel system (information and photograph submitted by Aquionics Inc.) (mgd ÷ 0.2642 = ML/d).

Picture		
Site background	Location	Gainesville, Georgia
	Plant process type	Oxidation ditch
	Use	River discharge/watershed
	Effluent limit	23 cfu/100 mL
	Procurement method	Public contractor bid
Key design parameters	Design dose	30 mJ/cm^2 average dose
	Design method	Computation fluid dynamic modeling
	Design (peak) flow	3783 m^3/h (24 mgd)
	Average flow	—
	UV transmittance	65%
	Total suspended solids	<15 mg/L
Equipment	Manufacturer	Aquionics
	Model	InLine+
	Type	Closed vessel—horizontal, perpendicular lamps
	Configuration	6 vessels—3 parallel trains
	Total number of lamps	96
	Redundancy	50%—1 train
	Power consumption per mgd	8.4 kW/mgd
	Headloss at design flow	43.2 cm (17 in.)
Installation and commissioning	Year commissioned	2001 and 2003
	Performance testing	Fecal coliform at commissioning
	Installation location	Indoors
Operational and control strategy	Control strategy	Intensity based
	Ballast turndown range	60–100%
	Level control method	Not applicable—pressurized vessel
Operation and maintenance	Annual operation	365 d/yr
	Lamp replacement period	8000 hours
	Ballast replacement period	20 years
	Sleeve cleaning method	Motor-driven automatic mechanical

and lower civil construction costs, especially when used with medium pressure lamps. However, for large systems, the stainless steel reactor cost can be prohibitive. Operational concerns related to entrapped air and algae buildup are typically cited as negative design aspects with these configurations; therefore, a sound understanding of upstream processes is necessary.

5.0 REFERENCE

U.S. Environmental Protection Agency (2006) *Ultraviolet Disinfection Guidance Manual for the Final Long Term 2 Enhanced Surface Water Treatment Rule*; EPA-815/R-06-007; U.S. Environmental Protection Agency: Washington, D.C.

Index

A

Absorbance, 20
Absorbance, dissolved constituents effects, 98
Acceptance, system, 166
Action spectrum, 25, 26
Air compressors/blowers, 185
Alarms, 175
Analysis, performance testing data, 165
Ancillary facilities, design, 151
Ancillary systems, 184
    air compressors/blowers, 185
    bank/module lifting, 185
    chemical cleaning tanks/stations, 184
Available head, 144

B

Bacteria, dose response, 34
Ballast cabinets, 150
Ballast replacement, 182
Ballasts and drivers, 135
Bank/module lifting, 185
Banks, system redundancy, 177
Bioassay methods
    existing data, 63
    log inactivation equation, 73
    microbiological testing, 61
    protocol for wastewater, 51, 53
    reporting, 64
    testing, 70
    validation, 65
Blowers, 185

C

Calculations, dose, 30, 32
Capital costs, 131
Case studies, 191
    closed vessel, 201
    open channel, horizontal flow, 191
    open channel, nonhorizontal flow, 197
Channel
    cleaning, 183
    cleaning tanks/stations, 184
    system redundancy, 177
Chemical hazards, 170
Cleaning components, 136
Closed vessel case study, 201
Closed vessel reactor, 114
Closed-channel systems, 144
Collimated beams, 30
    bioassay testing, 70
    dose measurements and calculations, 30, 32
    testing, 62
Commissioning startup and testing plan, 161
Commissioning, 153
Constructability and facility operations during construction, 141
Construction
    considerations, 158
    costs, 132
Construction management at risk, 155
Control cabinets/programmable logic controllers, 178
Control panels, 150
Costs
    ballasts and drivers, 135
    capital, 131
    cleaning components, 136
    construction, 132

intensity sensors and transmittance analyzers, 136
life cycle, 137
operation and maintenance, 133
operations and maintenance labor, 136
power consumption, 133
replacement parts, 134
Criteria, 4, 8

D
Data analysis, performance testing, 165
Deoxyribonucleic acid absorbance, 24, 26
Design criteria
flow, 116
headloss and water level, 117
influent and effluent water quality, 118
reactors, 116
Design dose and dose control strategies, 118, 119
Design
available head, 144
ballast cabinets, 150
closed-channel systems, 144
control panels, 150
facility, 129, 143
flow control, 145
flow distribution, 145
flow measurement, 145
flow splitting, 145
inlet hydraulics, 145
instrumentation, 150
level control, 146
open-channel systems, 144
power requirements and power redundancy, 146
process, 95
process redundancy, 147
sleeve cleaning methods and ancillary facilities, 151
ultraviolet system layout, 148

Design/bid/build, 155
Design/build, 156
Dissolved constituents, 98
Dissolved organic matter, 98
Divergence factor, 32
Dose calculations, 30, 32
divergence factor, 32
Petri factor, 32
reflection factor, 32
water factor, 32
Dose control, 118
Dose delivery, 61, 119
Dose requirements, 34
Dose responses
bacteria, 34
collimated beam testing, 30
microbes, 33
pathogens and surrogates, 30
protozoa, 35
viruses, 34
Drivers and ballasts, 135

E
Effluent water quality, 97
Electrical and control systems, 177
Electrical hazards, 169
Electromagnetic spectrum, 18
Energy conservation, 175
Energy costs, reactor validation, 90
Enforcement, manufacturer guarantees, 166
Equipment, 17
procurement methods, 156
replacement parts, 134
selection, 129
ultraviolet, 131
Excimer lamp, 40

F
Facility design, 129, 143
Flow
control, 145
design criteria, 116
distribution, 145

measurement, 145
performance testing, 164
splitting, 145
Flowrate monitoring, 173
Fouling of lamp sleeves, lamp racks, and channels, 113
Functional acceptance testing, 162

G
Germicidal action, 22
deoxyribonucleic acid and protein absorbance, 24
nucleic acid damage, 23
ultraviolet action spectrum, 25
Grotthuss–Draper law, 21
Guarantees, manufacturer, 166
Guidance, 11

H
Hazards
chemical, 170
electrical, 169
lifting, 170
other, 171
ultraviolet radiation, 169
Headloss
effects, 140
performance standards and system acceptance, 166
water level, design criteria, 117
History, 3
Hydraulics, 145
site and system, 143
troubleshooting, 187

I
Indicator bacteria, 5, 8
Influent and effluent water quality, design criteria, 118
Inlet hydraulics and flow splitting, 145
Innovations, 79
Inorganic compounds, 99
Installations, comparison, 143

Instrumentation, 150
Intensity sensors and ultraviolet transmittance analyzers, 136

L
Lamp
excimer, 40
fouling of sleeves and racks, 113
low-pressure high-output, 37
low-pressure, 36
medium-pressure, 37
microwave ultraviolet radiation, 38
orientation, 115
power measurement, 68
pulsed ultraviolet lamps, 39
replacement, 180
spacing, 115
technologies, 35, 37
testing, 59
variability and ultraviolet sensor port window testing, 58
Laws of photochemistry, 21
Layout
ballast cabinets, 150
control panels, 150
instrumentation, 150
ultraviolet system, 148
Level control, 146
Level monitoring, 174
Life cycle costs, 130, 137
Lifting devices, 149
Lifting hazards, 170
Light-emitting diodes, 38
Log inactivation equation, 73
Low ultraviolet transmittance, 185

M
Maintenance considerations, 180
ballast replacement, 182
channel cleaning, 183
lamp replacement, 180
sleeve cleaning, 182
sleeve replacement, 182

Manufacturer guarantees, performance standards and system acceptance, 166
Manufacturer reliability, 142
Mercury vapor lamps, 35–37
Microbe dose response, 33
    bacteria, 34
    protozoa, 35
    viruses, 34
Microbial repair and regrowth, 27
    dark repair, 29
    photo-reactivation, 28
    regrowth, 29
Microbiological testing, 61
Microwave ultraviolet radiation, 38

N
Non-submerged lamp systems, 115
Nucleic acid damage, 23

O
Open channel
    horizontal flow case study, 191
    nonhorizontal flow case study, 197
    reactor, 115
    systems, 144
Operation and maintenance, 142, 167
    costs, 133
    during construction, 141
    labor, 136
    safety, 168
    strategies, 175
Operational acceptance testing, 163

P
Particles, 100, 105
Pathogens, 30
Performance standards and system acceptance, 166
    enforcing manufacturer guarantees, 166
    headloss, 166
    power consumption, 166
Performance testing, 163
    alternative methods, 165
        stress testing, 165
        velocity profile testing, 165
    data analysis, 165
    flow, 164
    testing duration, 164
    testing protocols, 163
    water quality, 164
Permit requirements, 96
Petri factor, 32
Photochemistry, 21
Photon properties, 19
Power consumption and system efficiency, 133
Power consumption, performance standards and system acceptance, 166
Power measurement, lamp, 68
Power requirements and power redundancy, 146, 177
Process concepts, 17
Process design, 95
Process monitoring, 171
    flowrate monitoring, 173
    level monitoring, 174
    temperature monitoring, 175
    ultraviolet transmittance meters, 173
    ultraviolet-intensity sensors, 171
    warnings and alarms, 175
Process redundancy, 147
Programmable logic controllers, 178
Project delivery, 129, 153, 154
    construction management at risk, 155
    design/bid/build, 155
    design/build, 156
    equipment procurement methods, 156
    startup and commissioning, 153
Properties
    photon, 19
    ultraviolet light, 19

Protein absorbance, 24
Protozoa, dose response, 35

R
Radiation, ultraviolet, 169
Reactor analysis, 79
Reactor configurations, 114
    closed vessel reactor, 114
    lamp orientation, 115
    lamp spacing, 115
    non-submerged lamp systems, 115
    open channel reactor, 115
Reactor performance, 81
Reactor selection considerations, 114
Reactor validation, 79, 83
Reactor validation, energy costs, 90
Redundancy
    power, 146, 177
    system, 177
Reflection factor, 32
Regulatory considerations, 9, 10
Reliability, manufacturer, 142
Replacement parts, 134
    ballasts and drivers, 135
    lamps and sleeves, 134
Reporting, bioassay methods, 64

S
Safety, 168
    chemical hazards, 170
    electrical hazards, 169
    lifting hazards, 170
    other hazards, 171
    ultraviolet radiation hazards, 169
Scattering, 22
Service, 142
Site and system hydraulics, 143
Sizing example, 120
Sizing, system, 95
Sleeve cleaning methods and ancillary facilities, 151
Sleeve cleaning, 182
Sleeve replacement, 182
Sleeves, replacement parts, 134
Stark–Einstein law, 21
Startup
    checks, 162
    commissioning, 161
    functional acceptance testing, 162
    project, 153, 161
    testing, 161
Stochastic reactor validation, 91
Stress testing, 165
Supervisory control and data acquisition system integration, 178
System acceptance, 166
System redundancy (multiple banks and channels), 177

T
Temperature monitoring, 175
Testing
    bioassay, 70
    duration, 164
    microbiological, 61
    operational acceptance, 163
    performance, 163
    protocols, 163
    startup, 161
Total suspended solids, 103
Transmittance, 20, 103
    analyzers, 136
    chemical effects, 107
    dissolved constituents effects, 98
    industrial effects, 108
    low, troubleshooting, 185
    meters, 173
    seasonal and diurnal variability, 106
    secondary process effects, 106
    sidestream flow effects, 109
Troubleshooting, 185
    electrical issues, 185
    hydraulics, 187
    low ultraviolet transmittance, 185
    upstream process effects, 187
Turbidity, 103

U

Ultraviolet absorbance and transmittance, 20
Ultraviolet action spectrum, 25, 27
Ultraviolet light
    germicidal action, 22
    properties, 19
Ultraviolet radiation hazards, 169
Ultraviolet scattering, 22
Ultraviolet system layout, 148
    lifting devices, 149
    sleeve cleaning methods and ancillary facilities, 151
    ultraviolet system ballast cabinets, control panels, and system instrumentation, 150
Ultraviolet transmittance meters, 173
Ultraviolet-intensity sensors, 171
Upstream process effects, 187
Upstream processes, 110

V

Validation
    bioassay, 63, 65
    methods, 83
    protocols, 53
    reactor, 79
Velocity profile testing, 165
Viruses, dose response, 34

W

Warnings and alarms, 175
Warranties, service, and manufacturer reliability, 142
Water factor, 32
Water quality characterization tools, 101
    total suspended solids, 103
    transmittance, 103
        chemical effects, 107
        industrial effects, 108
        seasonal and diurnal variability, 106
        secondary process effects, 106
        sidestream flow effects, 109
    turbidity, 103
Water quality improvement, upstream processes, 110
Water quality, performance testing, 164

CPSIA information can be obtained
at www.ICGtesting.com
Printed in the USA
FSOW04n1303010616
21030FS